T0282344

Dehydroacetic Acid and Its Derivatives

Dehydroacetic Acid and Its Derivatives

Useful Synthons in Organic Synthesis

Edited by

Dr. Santhosh Penta M.Sc., Ph.D

Assistant Professor, Department of Chemistry, National Institute of Technology Raipur, Raipur, Chhattisgarh, India

elsevier.com

Elsevier
Radarweg 29, PO Box 211, 1000 AE Amsterdam, Netherlands
The Boulevard, Langford Lane, Kidlington, Oxford OX5 1GB, United Kingdom
50 Hampshire Street, 5th Floor, Cambridge, MA 02139, United States

Copyright © 2017 Elsevier Ltd. All rights reserved.

No part of this publication may be reproduced or transmitted in any form or by any means, electronic or mechanical, including photocopying, recording, or any information storage and retrieval system, without permission in writing from the publisher. Details on how to seek permission, further information about the Publisher's permissions policies and our arrangements with organizations such as the Copyright Clearance Center and the Copyright Licensing Agency, can be found at our website: www.elsevier.com/permissions.

This book and the individual contributions contained in it are protected under copyright by the Publisher (other than as may be noted herein).

Notices
Knowledge and best practice in this field are constantly changing. As new research and experience broaden our understanding, changes in research methods, professional practices, or medical treatment may become necessary.

Practitioners and researchers must always rely on their own experience and knowledge in evaluating and using any information, methods, compounds, or experiments described herein. In using such information or methods they should be mindful of their own safety and the safety of others, including parties for whom they have a professional responsibility.

To the fullest extent of the law, neither the Publisher nor the authors, contributors, or editors, assume any liability for any injury and/or damage to persons or property as a matter of products liability, negligence or otherwise, or from any use or operation of any methods, products, instructions, or ideas contained in the material herein.

British Library Cataloguing-in-Publication Data
A catalogue record for this book is available from the British Library

Library of Congress Cataloging-in-Publication Data
A catalog record for this book is available from the Library of Congress

ISBN: 978-0-08-101926-9

For Information on all Elsevier publications
visit our website at https://www.elsevier.com/books-and-journals

Working together
to grow libraries in
developing countries

www.elsevier.com • www.bookaid.org

Publisher: John Fedor
Acquisition Editor: Anneka Hess
Editorial Project Manager: Anneka Hess
Production Project Manager: Paul Prasad Chandramohan
Designer: Greg Harris

Typeset by MPS Limited, Chennai, India

Contents

2.   Ring Transformations in Dehydroacetic Acid

Santhosh Penta

3.   Synthesis of Different Heterocyclic Compounds by Using DHA

Gudala Satish

4. Dehydroacetic Acid–Metal Complexes

Gudala Satish

List of Contributors

Santhosh Penta National Institute of Technology Raipur, Raipur, Chhattisgarh, India

Gudala Satish National Institute of Technology Raipur, Raipur, Chhattisgarh, India

Archi Sharma National Institute of Technology Raipur, Raipur, Chhattisgarh, India

Chapter 1

Introduction

Santhosh Penta

National Institute of Technology Raipur, Raipur, Chhattisgarh, India

1.1 WHAT IS DEHYDROACETIC ACID?

Dehydroacetic acid (3-acetyl-4-hydroxy-6-methyl-2*H*-pyran-2-one) commercially abbreviated as **DHA** (**1**), is an important oxygen heterocyclic compound. It is well known for its various synthetic and physiological applications.

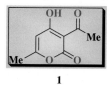

1

- Dehydroacetic acid is basically used as a bactericide and fungicide.
- It is utilized as a preserver to lessen pickle bloating for strawberries, squash, and different products.
- Sometimes, the salt form of dehydroacetic acid (known as sodium dehydroacetate) is used. This preservative shows excellent activity against fungi and has mild antibacterial properties.

It has been reported that due to its excellent chelating capacity in modern coordination chemistry and also its reactions, **DHA** and its derivatives have a wide utility in the field of organic synthesis. **DHA** is used in the formulation of a wide variety of products including skin care products [1] (Fig. 1.1), suntan, sunscreen lotions, fragrance, bath products, shaving gels, hair and nail care products, as well as eye and facial makeup.

It is obtained from both natural [2–4] and synthetic sources and is mainly used to produce clopidol (**2**). It is a coccidiostatic agent (Fig. 1.2) and a food preservative [5].

Since the present investigation deals with the studies on the synthesis of some potential dehydroacetic acid derivatives, it would be more appropriate and necessary to present here a brief review on the chemistry of dehydroacetic acid.

Dehydroacetic Acid and Its Derivatives. DOI: http://dx.doi.org/10.1016/B978-0-08-101926-9.00001-8
© 2017 Elsevier Ltd. All rights reserved.

FIGURE 1.1   Cosmetics. *Ying G, Kurunthachalam K. A survey of phthalates and parabens in personal care products from the United States and its implications for human exposure. Environ Sci Technol 2013;47:14442 – 9.*

(A)   (B)

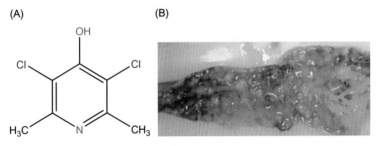

FIGURE 1.2   (A) Clopidol. (B) Coccidiosis in a goat ileum. *Reza K, Saeid RN, Zeinab Y. J Parasit Dis, Indian Society for Parasitology. 2012. doi:10.1007/s12639-012-0186-0.*

1.2   CHEMISTRY OF DEHYDROACETIC ACID

1.2.1   Constitution and Structure

DHA was discovered [6] in 1866 but ever since the discovery, the structure of **DHA** was a matter of great controversy. Two structures (**1** and **3**) were proposed by Fiest [7] and Collie [8].

Its structure was resolved by Rassweiler and Adams [9] in 1924. They obtained the evidence to support Fiest's structure and disprove the proposal given by Collie. **DHA** exists [10] as acetyl methyl pyrandione (**4**) in the solid state. It can exist in any of five tautomeric forms **4–8** [11] in the solution state which are represented below.

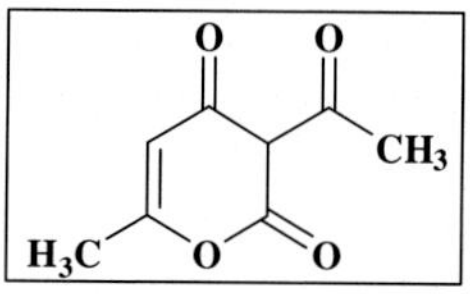

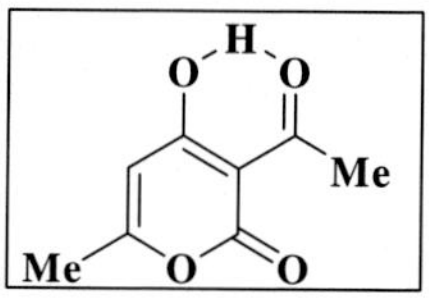

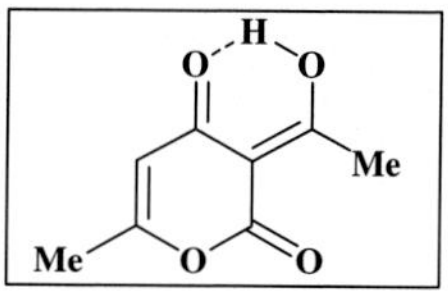

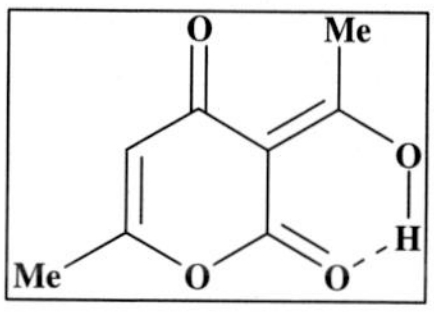

Finally, it exists in two tautomeric forms, i.e., 3-acetyl-4-hydroxy-6-methyl-2H-pyran-2-one (**5a**) and 3-acetyl-2-hydroxy-6-methyl-4H-pyran-4-one (**5b**), due to a quick tautomeric equilibrium between forms **5** and **6**, which are represented in Scheme 1.1.

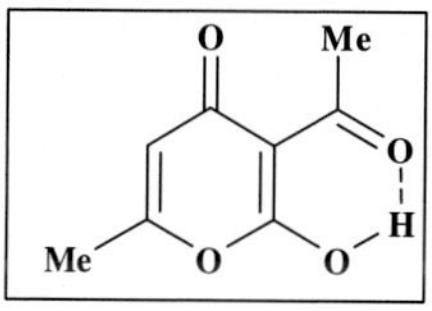

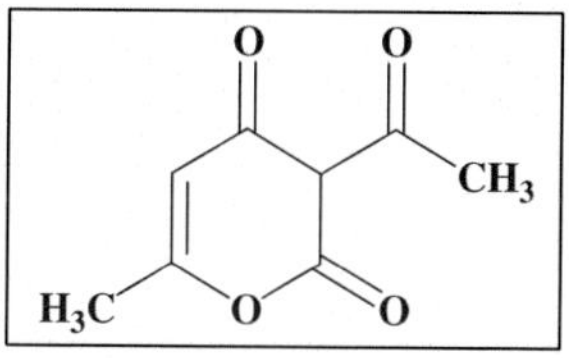

Based on its reactions and synthesis, its constitution has been proven. The 4-hydroxy structure (**5a**) has been assigned [12] based on its enolic character which was evidenced by a positive $FeCl_3$ test, analogy, data from the Diels-Alder

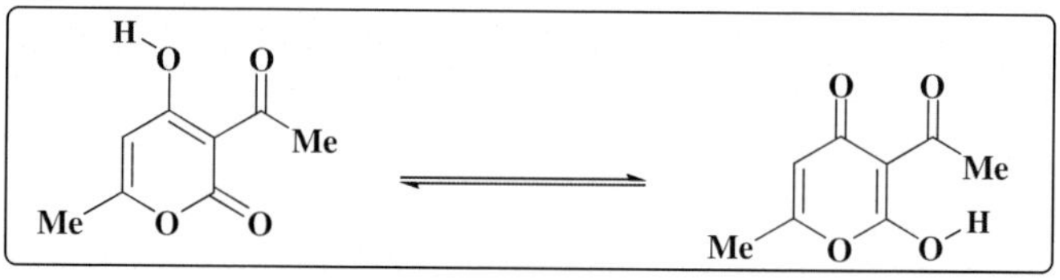

SCHEME 1.1   Tautomeric forms of DHA

TABLE 1.1 Spectral Data of DHA

UV	λ max (EtOH) 307 (logε 4.02), 227 nm (4.08), 207 (3.91)
IR (KBr, cm^{-1})	3400 (m, OH), 3030 (s, CH$_3$), 1710 (s, lactone C = O), 1640 (s, COCH$_3$), 1260 (s, C–O enolic), 1000 (m, C–O–C)
^{1}H-NMR (80 MHz, CDCl$_3$, δ ppm)	2.27 (s, 3H, C$_6$-CH$_3$), 2.68 (s, 3H, COCH$_3$), 5.92(s, 1H, C$_5$-H), 16.71 (s, 1H, OH exchangeable with7 D$_2$O)
^{13}C-NMR (20 MHz, CDCl$_3$, δ ppm)	20.36 (q, 6-CH$_3$), 29.63 (q, COCH$_3$), 101.13 (d, C$_5$H), 99.59 (s, C-3), 160.77 (s, C-6), 168.89 (s, C-2), 180.84 (C-4), 204.91 (s, COCH$_3$)
Mass (70 eV) (%)	168 (100), 153 (63.5), 140 (4.3), 129 (3.5), 126 (3.5), 125 (33), 111 (15), 102 (3.5), 101 (2.6), 99 (10), 98 (5.7), 85 (80), 84 (17.4), 74 (3.5), 73 (2.6), 69 (29.6), 67 (4.3), 60 (2.6), 55 (9.6), 53 (6.1), 43 (93), 42 (10), 41 (3.5), 39 (9.6), 29 (13.8)

FIGURE 1.3 Johann Georg Anton Geuther. *Carl D, Kurt H, Anton GSL, Seine A, Beric der Deuts Chemi Gesell 1930:145–57.*

reaction, and also on the fact [13] that the H-atom at C-3 lies in close proximity to three adjacent C=O groups.

The structure of **DHA** was fully established later by using the data obtained from spectroscopic analysis [14–16] as mentioned in Table 1.1.

1.2.2 Preparation of DHA

Geuther (Fig. 1.3) was the scientist [17] who reported the first preparation method of **DHA** in 1866. Since then it has been prepared by the cyclization of

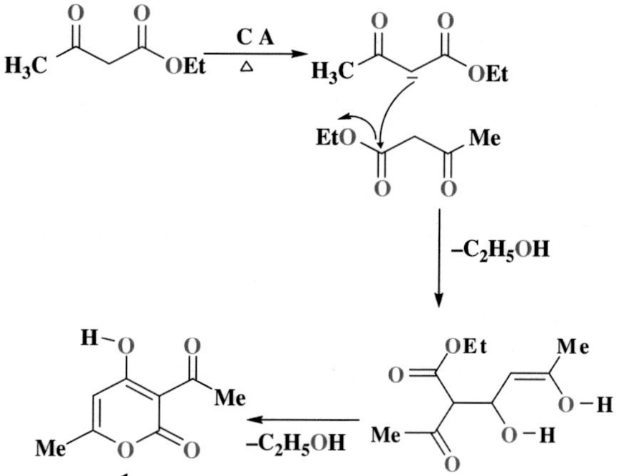

SCHEME 1.2 Synthesis of DHA

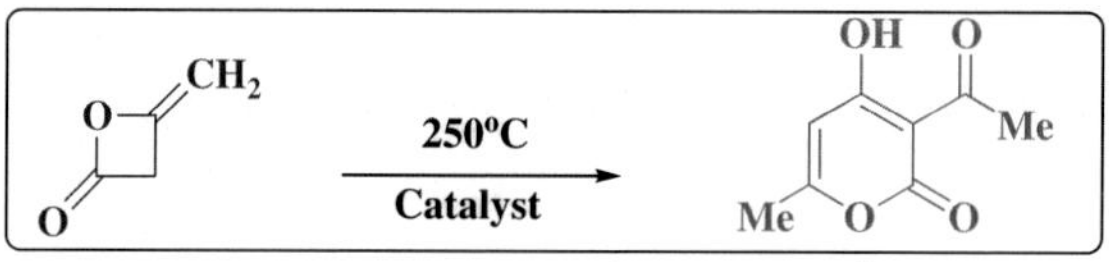

SCHEME 1.3 Formation of DHA from ketene

two equivalents of ethyl acetoacetate in the presence of a condensation agent. The success of the preparation strategy mainly depends on the ability of the condensing agent. The condensing agents are frequently found in literature and include NaHCO$_3$ [15,18], PhONa [19], Cu [20], Pb (OAc)$_4$ [21], Ag$_2$O, BaO, MgO, or amberlite [22], etc. (Scheme 1.2).

Balenovic had predicted the formation of **DHA** by the oxidation of ethyl acetoacetate with Pb (OAc) which was followed by the ketene as an intermediate [21] at room temperature.

In industry, **DHA** can be prepared by the dimerization of diketene using catalysts like NaOAc [23], pyridine [24], imidazole [25], NaOPh [26], etc. (Scheme 1.3).

Another notable procedure [27] for preparing **DHA** with 90%–94% yield involves the passing of a mixture of ethyl acetoacetate and nitrogen gas into a quartz tube filled with pumice and heated at 350–550°C.

1.3 REACTIVITY

DHA has four possible sites for attack [28] (Fig. 1.4), the lactone carbonyl at **2 position**, the carbonyl group of the acetyl side chain at **3 position**, the carbon atom of the potential carbonyl group at **4 position**, and the methyl carbon atom at **6 position**.

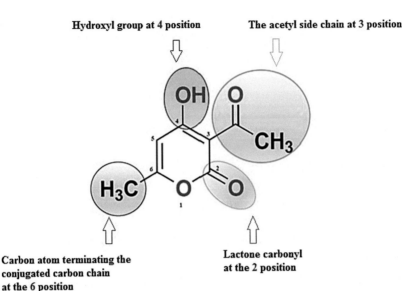

FIGURE 1.4 Reactive sites on DHA. *Stephen JF, Marcus E. Reactions of dehydroacetic acid and related pyrones with secondary amines. J Org Chem 1969;34:2527–34.*

FIGURE 1.5 Electrophilic and nucleophilic attacks on DHA. *Stephen JF, Marcus E. Reactions of dehydroacetic acid and related pyrones with secondary amines. J Org Chem 1969;34:2527–34.*

The carbonyl carbon atoms which are at C_2, C_3, C_4, and C_6 positions are strongly nucleophilic in nature and the C_3 and C_5 positions are electrophilic in nature (Fig. 1.5). Substitution reactions are more favorable at the C_4 position.

The attack of nucleophiles at C_2 and C_6 leads to the initial opening of the ring, which is followed by a different cyclization to form a new heterocyclic system. The C_5 position is quite inert with halogenation. The methyl group at C_6 and the acetyl group at C_3 positions are functionalized in different ways so as to confer electrophilic or nucleophilic reactivity.

1.4 REACTIONS AT C_3, C_4, C_5, AND C_6 POSITIONS OF DHA

1.4.1 Reactions of Carbonyl Group at C_3

1.4.1.1 Synthesis of Pyrazoles

The 3-cinnamoyl-2-pyrones (**9**) (chalcones) were formed by the condensation of **DHA** with aromatic aldehydes [29,30] in CHCl₃ in the presence of piperidine as

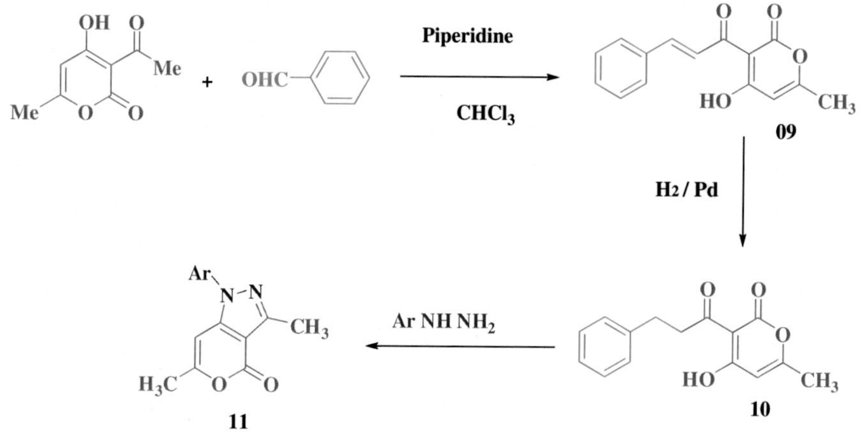

SCHEME 1.4 Synthesis of pyrazoles

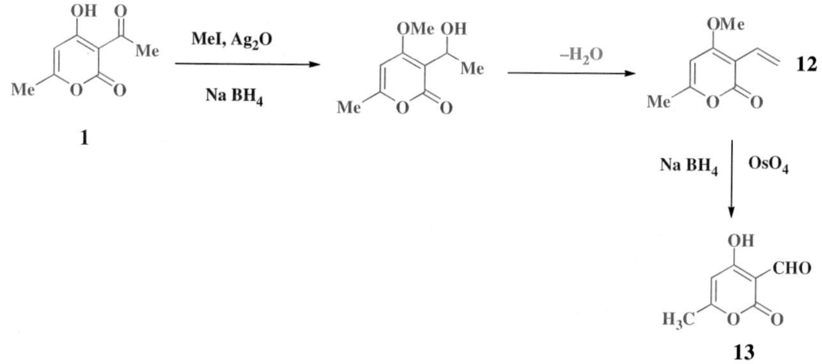

SCHEME 1.5 Formylation of DHA

catalyst. These chalcones undergo catalytic reduction [31] at the olefinic bond to form the corresponding dihydro derivative, namely 3β-arylpropionyl-2-pyrones (**10**). These dihydro derivatives on further condensation with various substituted phenyl hydrazines [32] form biologically important pyrazoles (**11**) (Scheme 1.4).

1.4.1.2 Formylation of DHA

The acetyl group of **DHA** at C_3 position was formylated [33] by modification of the side chain and the methyl ether (**12**) was oxidized to afford 3-formyl-4-methoxy-6-methyl-2-pyrone (**13**) (Scheme 1.5).

1.4.1.3 Synthesis of Benzo/Benzopyrano Quinolizinines

The condensation of 3,4-dihydroisoquinoline [34] with **DHA** and methylated by $(CH_3)_2SO_4$ in the presence of EtOH forms benzo quinolizinines (**14**) whereas benzopyranoquinolizine (**15**) is formed when refluxed in MeOH (Scheme 1.6).

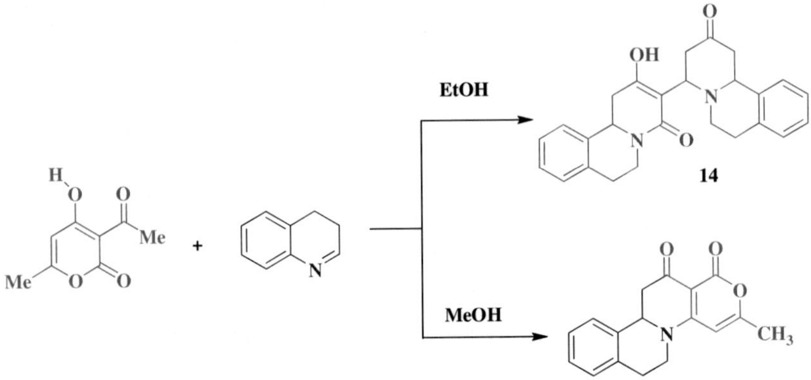

SCHEME 1.6 Synthesis of benzo/benzo pyranoquinolizinines

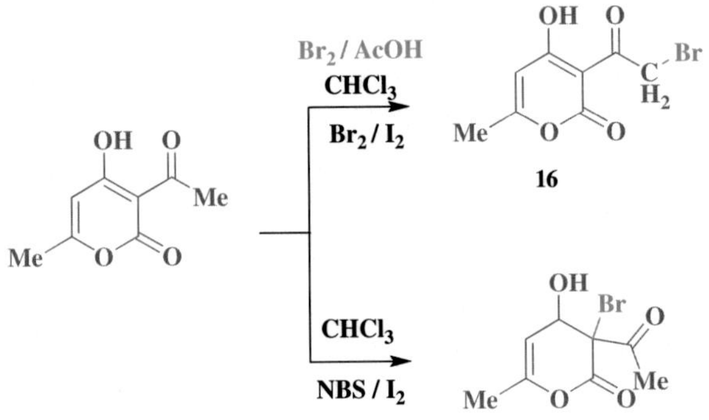

SCHEME 1.7 Halogenation of DHA

1.4.1.4 Halogenation

The acetyl group of the **DHA** at the C_3 position undergoes bromination [35,36] to afford acylbromopyrone (**16**) and also the carbon atom at the C_3 position undergoes bromination to give 3-bromo-pyrone (**17**) (Scheme 1.7).

1.4.1.5 Formation of Pyranopyran Diones

DHA on reaction [37] with ethylacetoacetate in the presence of piperidine and dilute HCl forms pyrano-2,5-pyrandione (**18**). During the synthesis of **DHA** using ethylacetoacetate, pyrano-4,5-pyrandione (**19**) also formed (Scheme 1.8).

1.4.1.6 Lengthening of Carbonyl Chain

DHA on reacting [37] with ethyl acetate using Na metal in dry ether affords pyranyl butanedione (**20**) (Scheme 1.9).

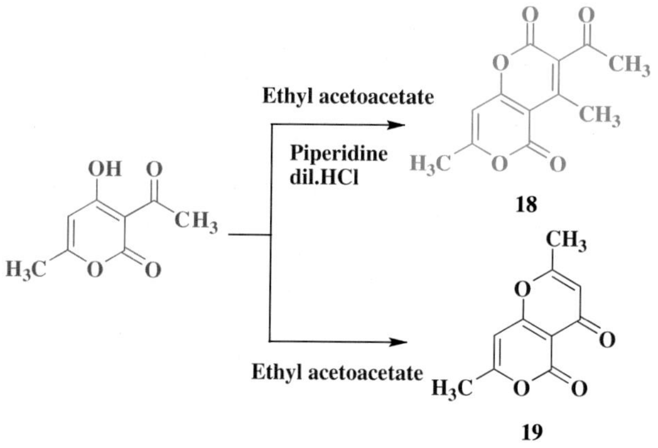

SCHEME 1.8   Synthesis of pyrano pyrandiones

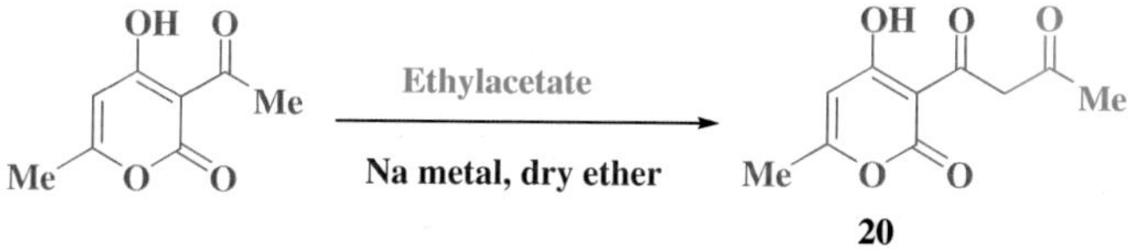

SCHEME 1.9   Synthesis of pyranyl butanediones

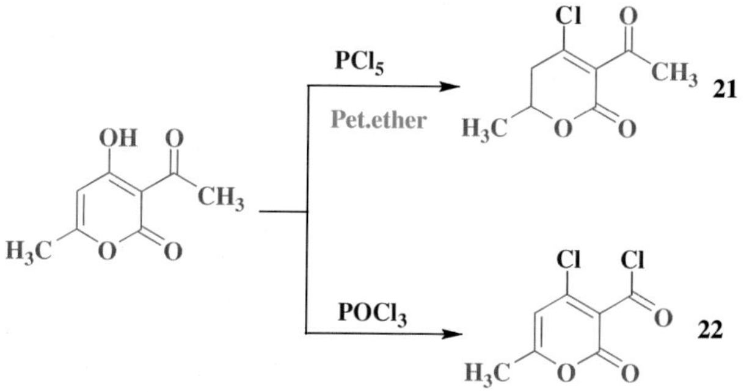

SCHEME 1.10   Synthesis of mono and dichloro derivatives of DHA

1.4.2   Reactions at C₄

1.4.2.1   Reactions With Nucleophiles

1.4.2.1.1   Synthesis of Mono and Dichloro Derivatives

DHA on chlorination with PCl₅ in the presence of ether forms monochloride (**21**), while with POCl₃, it forms dichloride (**22**) (Scheme 1.10).

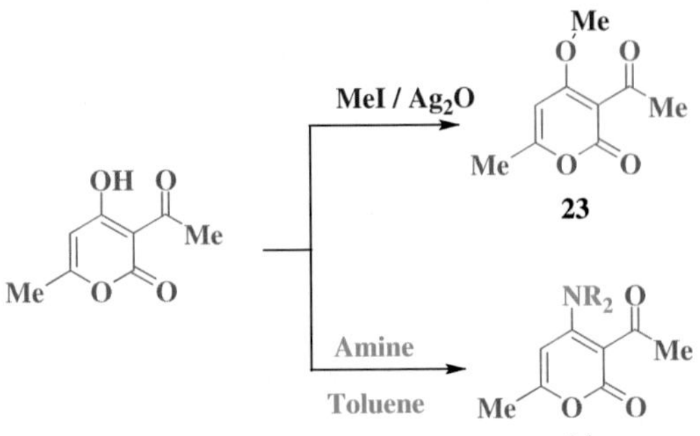

SCHEME 1.11 Synthesis of amino pyrones and methyl ethers

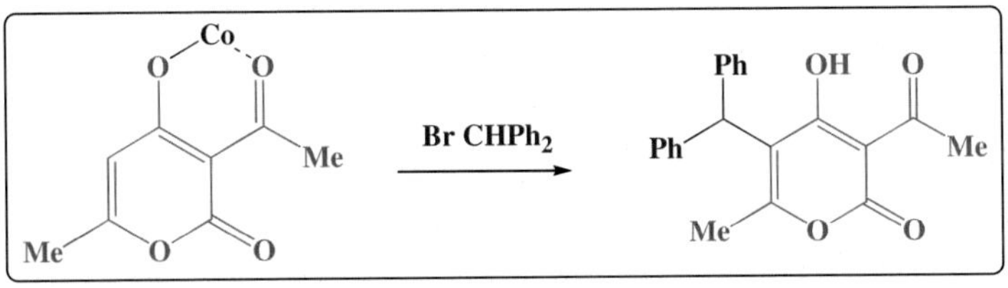

SCHEME 1.12 Synthesis of benzhydryl pyran-2-ones

1.4.2.1.2 Synthesis of Amino Pyrones and Methyl Ethers

DHA on reaction with methyl iodide of silver oxide forms methyl ethers (**23**), on the other hand, **DHA** with primary amines [38] and secondary amines in the presence of toluene gives amino pyrones (**24**) (Scheme 1.11).

This reaction reveals that the presence of a carbonyl group at the C_3 position enables the displacement of oxygen-based leaving groups.

1.4.3 Reactions at C_5

1.4.3.1 Synthesis of Benzhydryl Pyran-2-ones

The C_5 position of **DHA** is not sufficiently activated toward electrophile attack. The cobalt (II) complex of **DHA** (**25**) reacts with benzhydryl bromide to give pyrone (**26**). This reaction is the exception rather than general and it seems to involve free radicals [39,40] (Scheme 1.12).

1.4.3.2 Synthesis of Bromo Pyrones

DHA undergoes bromination at the C_5 position in the presence of bromine and chloroform to afford bromo pyrone (**27**). This reaction probably proceeds by the addition of bromine to the double bond at C_5–C_6 followed by the elimination

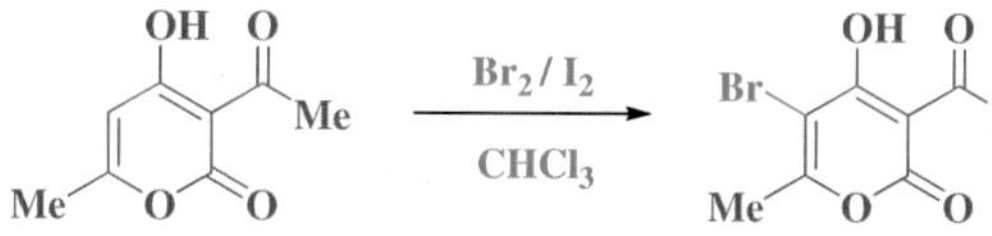

SCHEME 1.13   Synthesis of bromopyrones

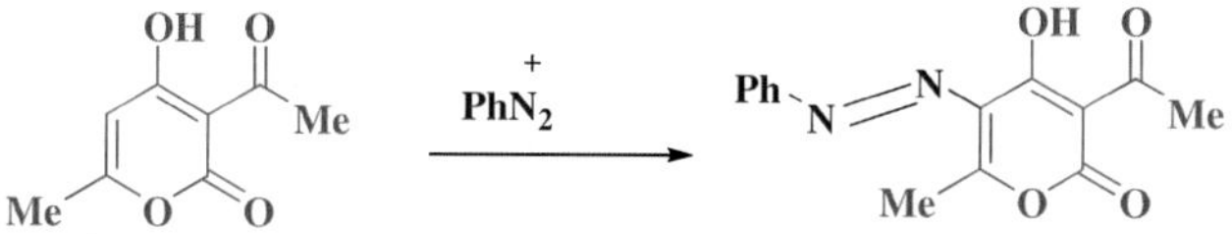

SCHEME 1.14   Synthesis of diazo compounds

of hydrogen bromide. This bromo pyrone (**29**) has permitted the preparation of a large number of pyrones with different functionalization in the rest of the molecule by transformation (Scheme 1.13).

1.4.3.3 Synthesis of Diazo Compounds

DHA reacts with diazonium cations to afford diazo compounds (**28**) due to the inertness of the C_5 position toward electrophiles (Scheme 1.14).

1.4.4 Reactions at C_6 Position

The methyl group at the C_6 position exhibits the behavior of allylic positions and can undergo halogenation under radical conditions and oxidation in the presence of appropriate reagents. The allylic bromination of **DHA** leads to a bromo derivative (**29**), which can be hydrolyzed to alcohol [41] (**30**) and transformed into the phosphonium bromide (**31**) and thioproducts [42] (**32**) (Scheme 1.15).

1.4.4.1 Synthesis of Substituted Compounds of DHA at C_6 Position

DHA on treatment with three equivalents of strong base, such as $NaNH_2$ in liquid ammonia, produces the corresponding trianion, which on quenching with one equivalent of electrophile, affords products [43] (**33**) arising from regioselective reaction at the most nucleophilic carbanionic center, the methyl group at C_6 (Scheme 1.16).

Electrophiles such as alkyl halides, benzophenone, and methyl benzoate were used in the reaction leading to the respective final products. This technique has been successfully applied particularly in the synthesis of pheromones

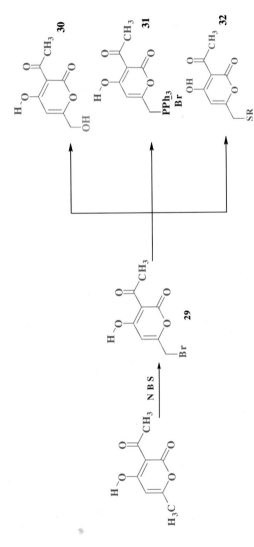

SCHEME 1.15   Synthesis of Alcohols, Phosphonium bromides and Thioproducts

SCHEME 1.16   Reactions on methyl group at C$_6$

FIGURE 1.6   Orientation of male moths toward the female-emitted sex pheromone (blue) in a natural environment. *Nina D, Fabienne D, Sylvia A, Michel R. Responses to pheromones in a complex odor world: sensory processing and behavior. Insects 2014;5:399–422.*

(a chemical produced by animals, humans, ants, and insects to impact the behavior of another animal, also known as behavior altering agents) [44–46] (Fig. 1.6).

1.4.4.2   Synthesis of Unsaturated Pyrones

Benzophenone undergoes aldol type condensation and affords carbinol pyrone (**34**) which on dehydration gives unsaturated pyrone (**35**) (Scheme 1.17).

1.5   OTHER REACTIONS

1.5.1   Hydrogenation of the Ring

DHA under controlled hydrogenation [47,48] and catalysis with palladium forms dihydropyrone (**36**) by selective introduction of hydrogen at the double bond C_5–C_6 along with a ketone of the carbonyl group at the C_3 position (Scheme 1.18).

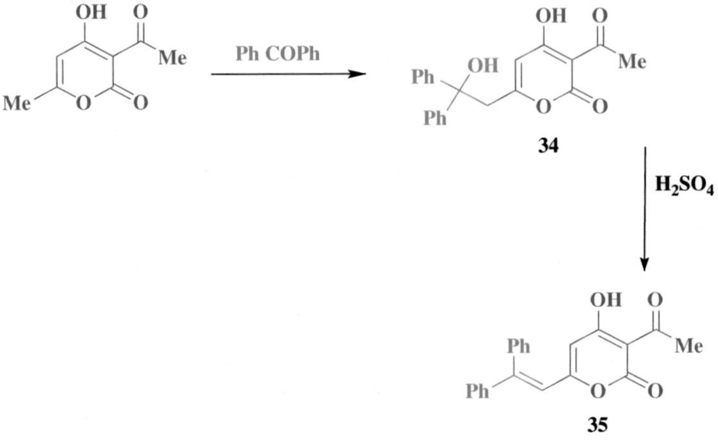

SCHEME 1.17 Synthesis of unsaturated pyrone

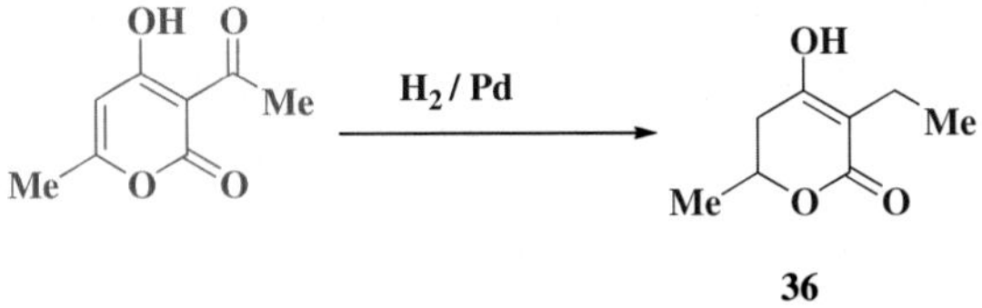

SCHEME 1.18 Synthesis of dihydropyrones

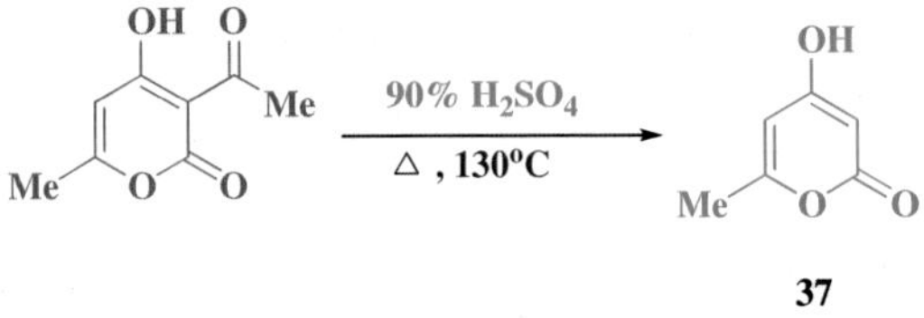

SCHEME 1.19 Deacetylation of DHA

1.5.2 Deacetylation

DHA on heating in the presence of 90% H_2SO_4 at 130°C produces 4-hydroxy-6-methyl-2-pyrone [49,50] (**37**) (Scheme 1.19).

1.5.3 Transformation Into Carbocyclic Systems

DHA on reacting [51] with sodium hydroxide forms a six-membered carbocycle, namely, orcinol (**38**) (Scheme 1.20). This reaction is an example of decarboxylative aldol condensation. The compound (**38**) obtained by the cyclization process sometimes mimic the biogenetic synthesis of phenolic compounds.

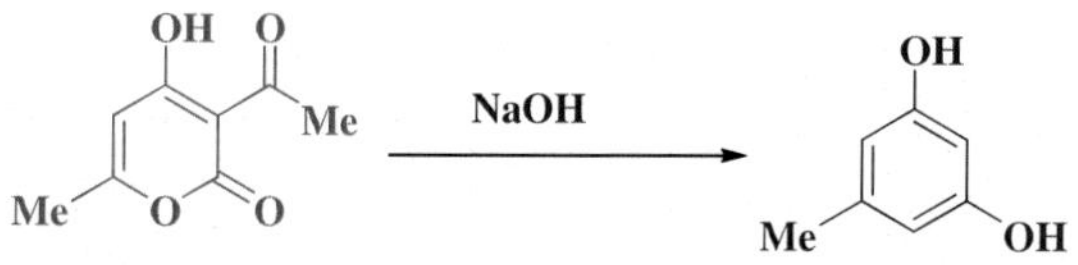

SCHEME 1.20   Synthesis of Orcinol

1.5.4   Formation of Metal Complexes

DHA and its Schiff bases are the best chelating agents and the general structure (**39**) of many complexes have been synthesized [52–55].

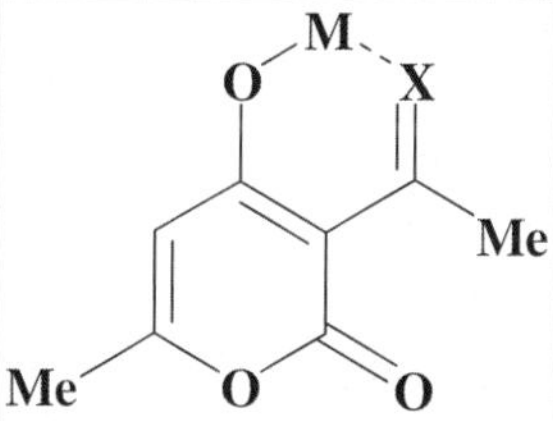

1.6   APPLICATIONS

DHA and its derivatives find an immense number of applications in various fields due to their high chemical reactivity and physiological properties.

1.6.1   Synthetic

DHA and its derivatives are a rich source of several types of heterocyclic compounds. 3β-aryl-propionyl-2-pyrones (**10**) are used as intermediates for the preparation of a series of biologically important pyrazoles (**11**). The acyl bromo pyrone (**16**) obtained from **DHA** is used in the dye-stuff industry. **DHA** is used as a synthon for preparing various pyrandiones (**18, 19**). The deacetylated product of **DHA** is hydroxy methyl pyrone (**37**) which can be used as a starting material for synthesizing many pyrones. Orcinol (**38**) is a compound obtained from **DHA** which can mimic the biogenetic synthesis of phenolic compounds.

1.6.2   Biological

DHA and its derivatives have numerous biological applications and among them a few are mentioned here. The copper salt of **DHA** [56] is used for the

(A)   (B)

FIGURE 1.7 (A) Sheath blight on rice plant. (B) Leaf blight on rice plant. *(A) Vijay Krishna Kumar K, Reddy MS, Kloepper JW, Lawrence KS, Zhou XG, Groth DE, et al. Commercial potential of microbial inoculants for sheath blight management and yield enhancement of rice. 2011: 237–63. doi:10.1007/978-3-642-18357-7_9 [Chapter 9]. (B) Barbara V. Underground life for rice foe. Nature 2004:431:516–7.*

control of bacterial and fungal diseases like sheath blight and leaf blight on rice plants [57,58] (Fig. 1.7).

The **DHA** anilino derivatives are used as an insecticide [59] and acaricide. In agriculture, phacidin [60] (**40**) and podoblastin [61] (**41**) can be used as fungicides.

40   **41**

Another class of pyrones, citreo viridines, are inhibitors of ATP synthesis [62]. These citreo viridines (**42**) are related to a disease, namely cardiac beriberi, an illness associated with yellowish rice in Asian countries.

42

On the other hand, 3-(pyridyl carbonyl) 2,4-pyrandiones are used as herbicide [63]. The potassium salt of **DHA** with isonicotinyl hydrazide shows antituberculosis activity in the long-term chemotherapy of clinical tuberculosis [64]. The cobalt complex of 3β-hydroxyphenyl propionyl hydroxymethyl pyrone

(**10**) exhibits potent anticancer activity [65]. The compound called Elasnin (**43**) isolated from *Streptomyces* is a specific inhibitor of an enzyme called elastase which is involved in inflammatory pulmonary emphysema [66] (a disease of the lungs that can make breathing difficult).

43

Hymenoquinone (**44**) is a pyrone extracted from piper methysticum roots which exhibits stimulating properties [67]. A few derivatives of **DHA** are used as anticoagulants [68]. Some metal chelates of **DHA** have also been studied for fungicidal screening [52,53,55].

44

1.6.3 Industrial

DHA and its derivatives are widely used to form various industrially applicable compounds. The **DHA** derivative called Clopidol (**2**) is a coccidiostatic agent which is used in the poultry industry. 3-acyl bromopyron [36] (**16**) is used as an intermediate in the dye-stuff and pharmaceutical industries. Instead of sodium benzoate, **DHA** and its sodium salts are used as good preservatives in the food and beverage industry. In plastic and rubber industries, some **DHA** metal complexes are used as heat stabilizers [69,70]. Copper gluconate of **DHA** sodium salt is used as z deodorant [71] in the cosmetic industry.

1.6.4 Other Allied Uses

DHA and its derivatives, in particular azo-methine or Schiff bases, have been well studied [52–55] for their complexation and as excellent chelating agents (bi-, tri-, and polydentate) to various transition metal ions to form stable metal complexes of novel structural interest.

Chapter 2

Ring Transformations in Dehydroacetic Acid

Santhosh Penta

National Institute of Technology Raipur, Raipur, Chhattisgarh, India

2.1 SYNTHESIS OF VARIOUS PYRAZOLES

2.1.1 Synthesis of Amino Phenyl Ethanimidoyl Methyl Phenyl Pyrazololes

The reaction of phenyl hydrazine with dehydroacetic acid (**1**) gives acetoacetylpyrazole as a key intermediate (**45**) (Scheme 2.1).

Further, this acetoacetylpyrazole (**45**) reacts with *o*-phenylenediamine (**46**) and reflux with an excess amount of ethanol gives 4-[(1*E*)-*N*-(2-aminophenyl) ethanimidoyl]-3-methyl-1-phenyl-1*H*-pyrazol-5-ols (**47**) (Scheme 2.2).

It has been noticed that heating of compound (**47**) in the presence of xylene prompts the development of important compounds 2-methyl-1*H*-benzo[d]imidazole (**48**) and methylphenylpyrazolole (**49**) (Scheme 2.3).

2.1.2 Synthesis of Hydroxy Phenyl Ethanimidoyl Methyl Phenyl Pyrazololes

As mentioned in the previous scheme, acetoacetylpyrazole (**45**) can be formed by the reaction of phenyl hydrazine with dehydroacetic acid. The reaction of *o*-amino-phenol (**50**) with β-diketone (**45**) and on refluxing with an excess amount of ethanol gives pyrazololes (**51**) (Scheme 2.4).

In contrast, when the compound (**51**) is heated under similar conditions as compound (**47**) it remains unchanged.

2.1.3 Synthesis of 4-Pyrazolylpyrimidobenzimidazoles

The reaction of acetoacetylpyrazole (**45**) with 2-aminobenzimidazole (**52**) and on refluxing with butanol gives 4-pyrazolylpyrimidobenzimidazole (**53**) (Scheme 2.5).

The mechanism of this reaction is mentioned in Scheme 2.6.

Dehydroacetic Acid and Its Derivatives. DOI: http://dx.doi.org/10.1016/B978-0-08-101926-9.00002-X
© 2017 Elsevier Ltd. All rights reserved.

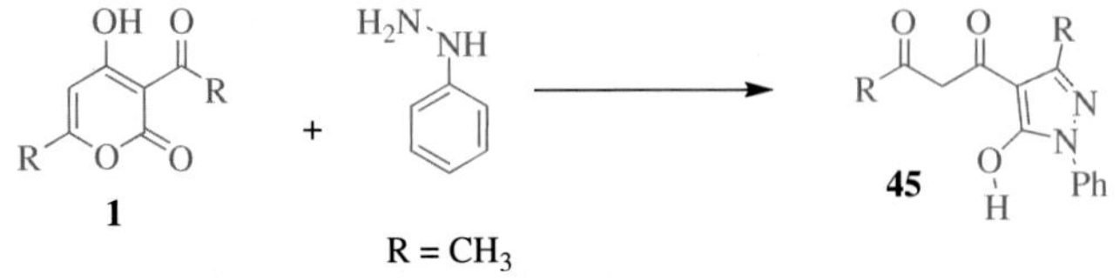

SCHEME 2.1 Synthesis of amino phenyl-ethanimidoyl-methyl phenyl pyrazololes.

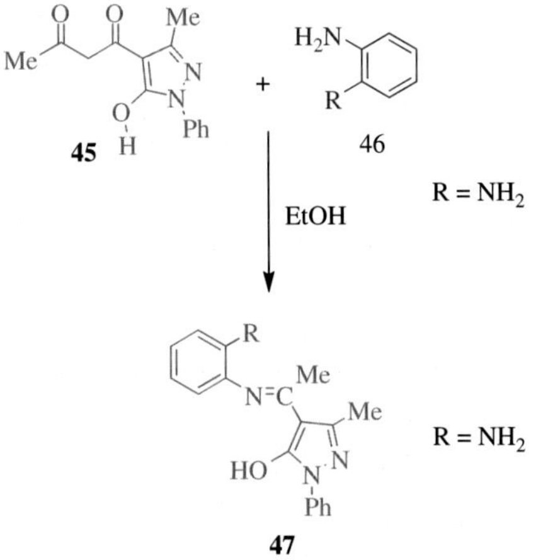

SCHEME 2.2 Synthesis of 2-aminophenyl ethanimidoyl-3-methyl phenyl-1H-pyrazol-5-ols.

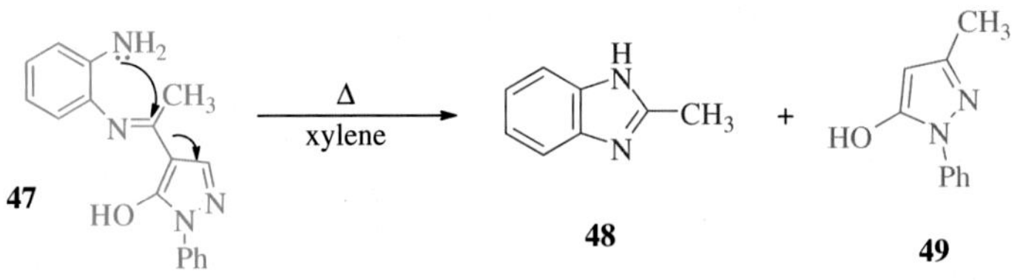

SCHEME 2.3 Synthesis of benzoimidazoles and methyl phenyl pyrazololes.

2.1.4 Synthesis of Pyrazolones, Pyridopyrimidinone, Methylphenylpyrazololes

The reaction of 2-aminopyridin-3-ol (**54**) with β-diketone (**45**) and on refluxing with ethanol affords a mixture of three products. From that, hydroxymethyl pyridopyrimidine ylidene methylphenylpyrazolone (**55**) was initially isolated then later hydroxymethylpyridopyrimidinone (**56**) and methylphenylpyrazolole (**57**) were isolated (Scheme 2.7).

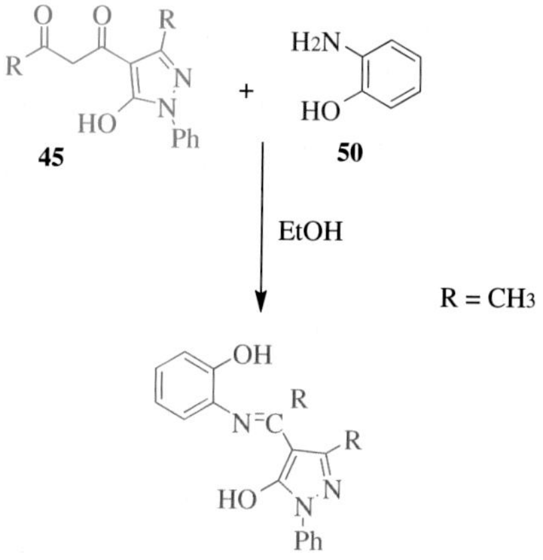

45 50 51

SCHEME 2.4 Synthesis of pyrazololes from o-amino-phenol and β-diketone.

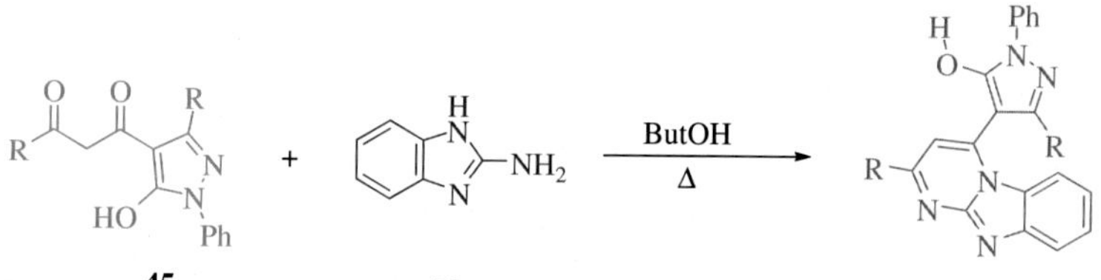

45 52 53

$R = CH_3$

SCHEME 2.5 Synthesis of 4-pyrazolyl pyrimido benzimidazoles.

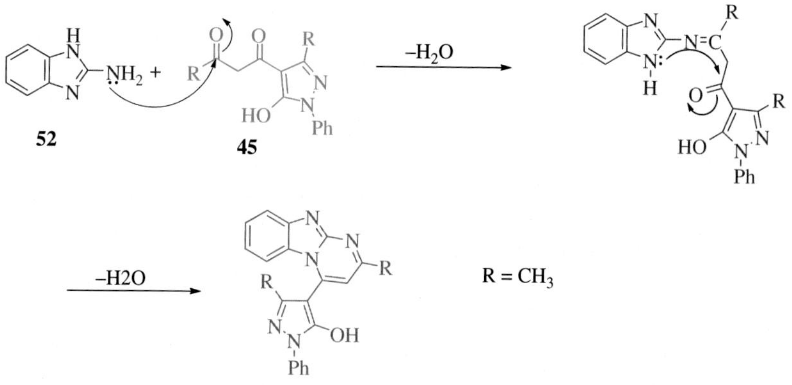

52 45 53

SCHEME 2.6 Mechanism of synthesis of 4-pyrazolyl pyrimido benzimidazoles.

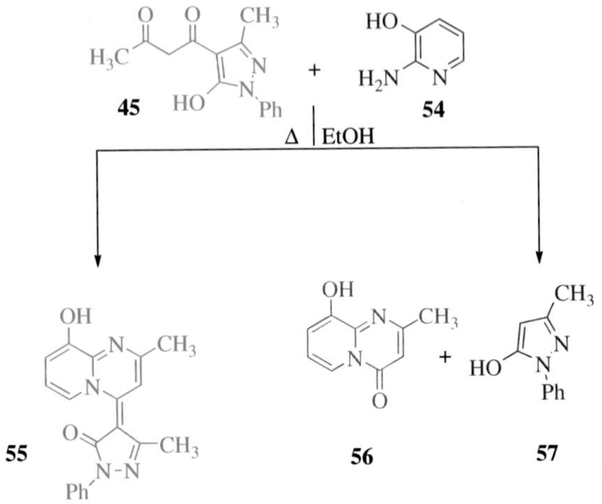

SCHEME 2.7   Synthesis of pyrazolones, pyridopyrimidinone, methyl phenyl pyrazololes.

The formation of **56** and **57** can be explained by the following mechanism. In the first step, **intermediate [A]** will be formed by the binding of the amino group of compound (**54**) on the carbonyl group acetyl side chain of compound (**45**). Then, it undergoes intramolecular cyclization by dehydration which leads to the formation of **intermediate [B]**. Later, it undergoes the attack of the endocyclic NH group of compound (**54**) and affords pyrazolones (**55**) (**path a**), pyridopyrimidinone (**56**) and pyrazololes (**57**) (**path b**) as shown in Scheme 2.8.

2.1.5   Synthesis of Pyrazolylidenethiazolopyrimidines

The compound (**45**) undergoes reaction with 2-aminothiazole (**58**) and on refluxing with butanol for 1 hour produces (*E*)-3-methyl-1-phenyl-4-(5*H*-thiazolo[3,2-a]pyrimidin-5-ylidene)-1*H*-pyrazol-5(4*H*)-ones (**59**) (Scheme 2.9).

2.1.6   Synthesis of Pyrazolylpyridones

According to the Guarreschi reaction, the reaction of compound (**45**) with 2-cyanoacetamide (**60**) and refluxion with ethanol and piperidine produces pyrazolylpyridones (**61**) in good yields (Scheme 2.10).

2.2   SYNTHESIS OF BIPYRAZOLES

In the synthesis if bipyrazoles, initially, 2-(4,6-dimethylpyrimidine-2-yl)-1-[1-(4-hydroxy-6-methylpyran-2-one-3-yl)ethylidene]hydrazines (**63**) can be synthesized by the reaction of **DHA** (**1**) with hydrazinodimethylpyrimidine (**62**) in the presence of ethyl alcohol. Further, it will be refluxed with the acetic acid

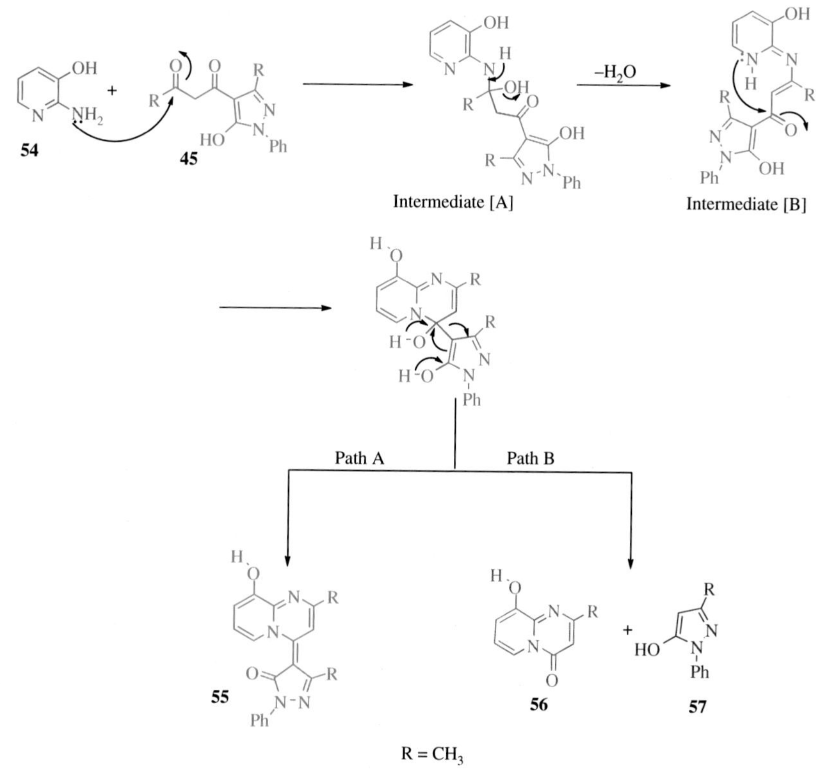

Intermediate [A]

Intermediate [B]

Path A | **Path B**

55

56 57

R = CH₃

SCHEME 2.8 Mechanism of Synthesis of pyrazolones, pyridopyrimidinone, methylphenylpyrazololes.

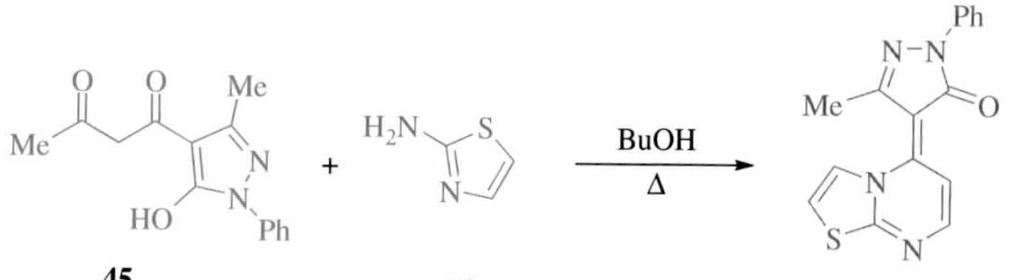

45

58

59

SCHEME 2.9 Synthesis of pyrazolylidene thiazolo pyrimidines.

to give the rearranged key intermediate, namely 5-hydroxy-3-methyl-1-(4,6-dimethylpyrimidin-2-yl) pyrazol-4-yl-1,3-butanedione (**64**). Lastly, the reaction between (**64**) and various aryl/heteroaryl hydrazines (**65**) in the presence of ethyl alcohol/acetic acid/sodium acetate [72] leads to the high yield of bipyrazoles (**66**) (Scheme 2.11).

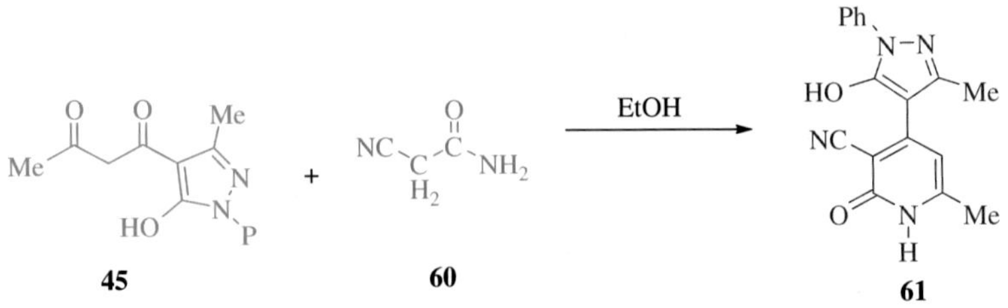

SCHEME 2.10 Synthesis of pyrazolyl pyridines.

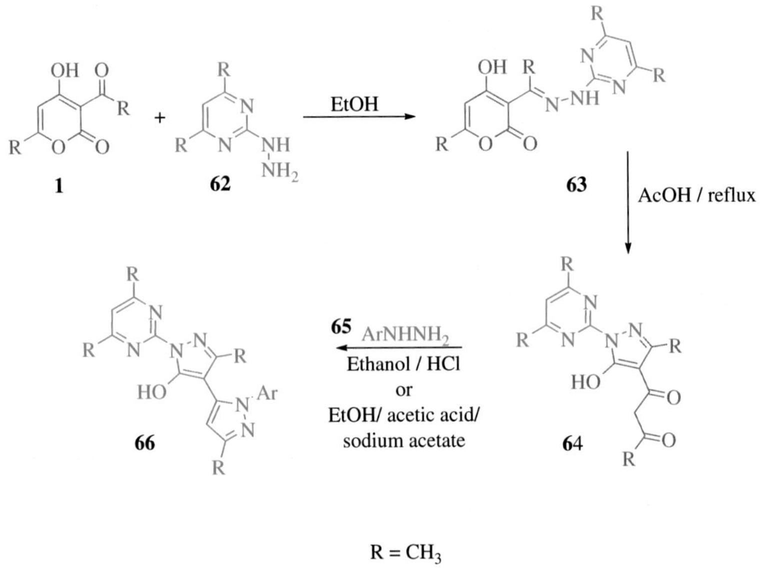

$$R = CH_3$$

SCHEME 2.11 Synthesis of bipyrazoles.

The derivatives of the these bipyrazoles are given in the following scheme.

The possible mechanism for this reaction is mentioned in Scheme 2.12. The amino group of hydrazine (**62**) shows a nucleophilic attack on the carbonyl group (>C=O) of the acetyl side chain which is present at the **3rd position** of **DHA** (**1**) to give hydrazones (**63**) by losing one water molecule.

Then the resulting hydrazine (**63**) undergoes a rearrangement reaction and generates the compound (**64**) by a nucleophilic attack of nitrogen on the lacto-carbonyl group of the **DHA** ring. In the last step, bipyrazolols (**66**) are formed by the condensation of pyrazolyldiketone (**64**) with binucleophilic aryl/heter-oarylhydrazines (**65**).

The reactions of dehydroacetic acid and its derivatives with hydrazines or hydroxylamine are a rich source for forming many kinds of heterocyclic systems. **DHA** (**1**) undergoes reaction with hydrazines or hydroxylamine to form

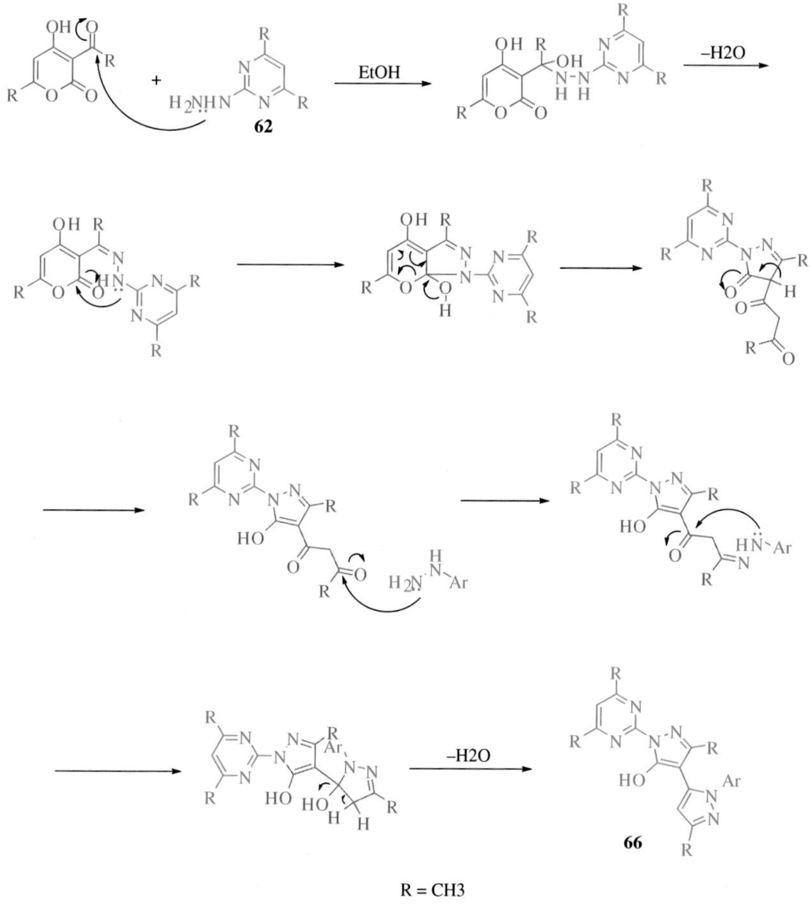

R = CH3

SCHEME 2.12 Mechanism of synthesis of bipyrazoles.

oxime (**67**) or hydrazones (**68**). The initially formed compounds (**67**) or (**68**) will regenerate into isoxazolyl 1,3-diones (**69**) or pyrazolyl 1,3-diones (**70**) by dehydroacetic acid ring cleavage, then these finally by substitution form biisox-azols [73] (**71**) and bipyrazols (**72**) (Scheme 2.13).

Substituted pyrazolylpyrazoles (**76**) can be synthesized in two ways. Initially, the **DHA** upon treatment with two equivalents of phenyl hydrazine hydrochlo-ride under refluxion conditions with aqueous HCl in DME includes the forma-tion of pyrazolyl 1,3-dione (**74**) as an intermediate followed by refluxion of hydrazone (**73**). In the second step, the initially formed pyrazolyl 1,3-dione (**74**) again reacts with one equivalent of phenyl hydrazine hydrochloride and finally forms pyrazolylpyrazoles (**76**) along with (**77**) and (**78**) (**path b**).

Bipyrazoles (**79**) are not formed by this mechanism, as mentioned in the Scheme 2.14 (**path a**). On the other hand, pyrazolyl 1,3-dione (**74**) involves

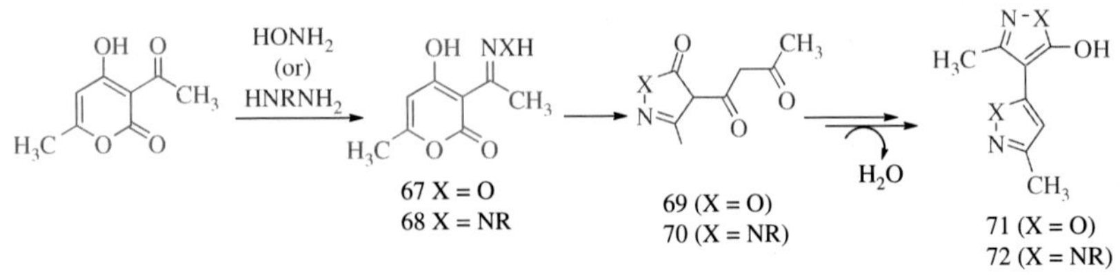

SCHEME 2.13 Synthesis of biisoxazols and bipyrazols.

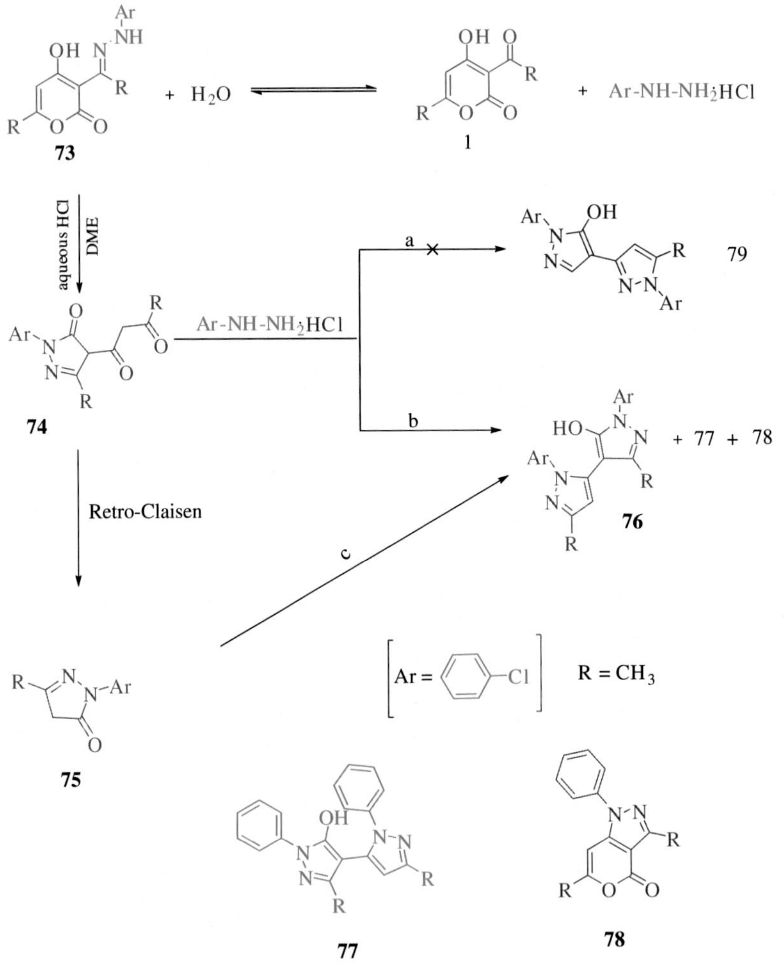

SCHEME 2.14 Mechanism of synthesis of biisoxazols and bipyrazols.

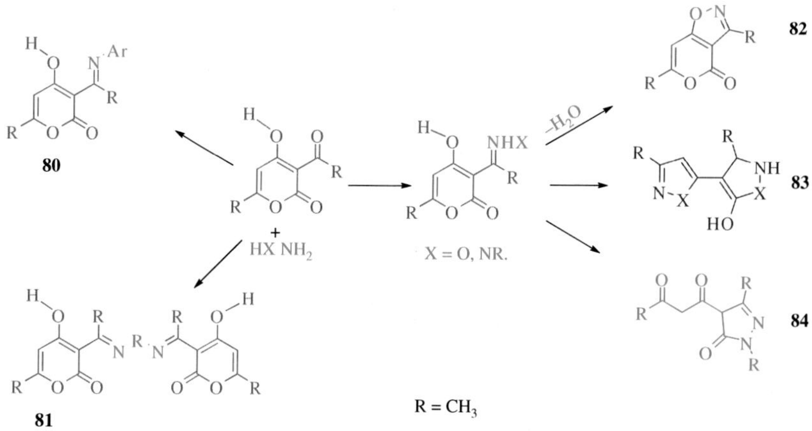

SCHEME 2.15 Synthesis of different heterocycles from DHA via ring opening, cyclization and transformation.

a simple retro-Claisen condensation (**path c**) and produces the closely related compound (**75**), which on heating yields compound (**76**).

2.3 ISOXAZOLES, PYRAZOLO-PYRAZOLES, AND PYRAZOLONES

DHA and its derivatives react with various amines to form several types of heterocyclics. They often result in ring-opening and cyclization, transforming into other heterocyclic systems. They form Schiff bases or azomethine derivatives (**80, 81**) by reacting with different mono- and diamines [74–78]. **DHA** on reacting with hydroxyl amines hydrazines gives isoxazole [73] (**82**), pyrazolo-pyrazoles [79] (**83**), and pyrazolones [80] (**84**) (Scheme 2.15).

2.4 SYNTHESIS OF PYRAZOLO-OXAZIN-2-THIONE COMPOUNDS

When **DHA** (**1**) reacts with phenyl hydrazine in the presence of acetic acid, it gives 1,3-dicarbonyl compound (**85**), then, it reacts with aliphatic and aromatic primary amines to form pyrazolyl-enaminones (**86**). In similar conditions, compound (**86**) reacts with thiophosgene in the presence of triethylamine to give N-substituted pyrazolo-oxazin-2-thione [81] compound (**87**) with 69%–86% yield (Scheme 2.16).

The mechanism of the synthesized N-substituted pyrazolo-oxazin-2-thione compounds is shown in Scheme 2.17.

2.5 SYNTHESIS OF PYRIDONES

There are a large number of illustrations for the transformation of dehydroacetic acid to pyridones. Here, **DHA** on reaction with ammonia and primary amines

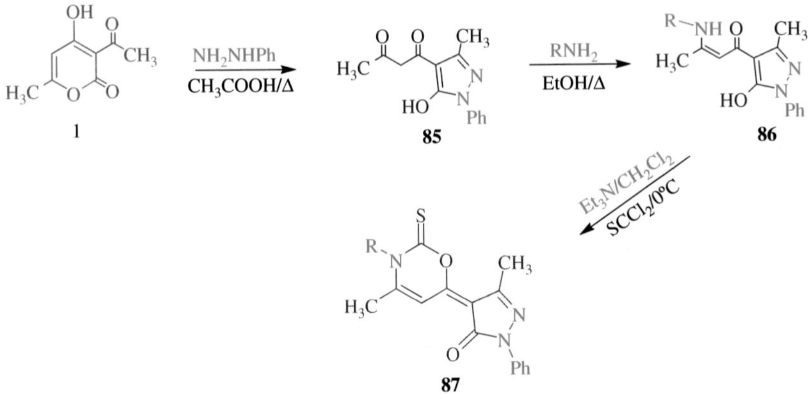

SCHEME 2.16   Synthesis of pyrazolo-oxazin-2-thione compounds.

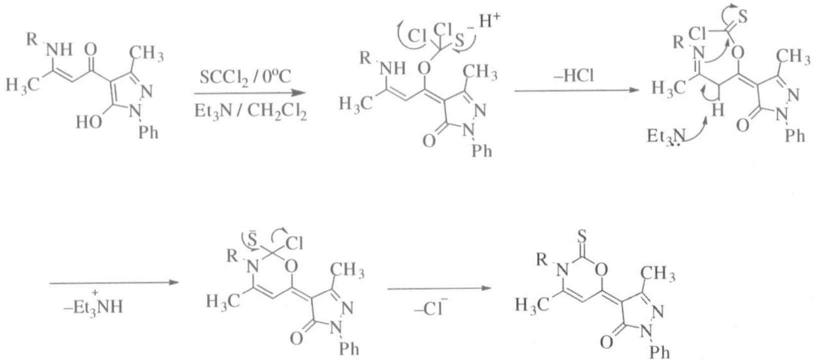

SCHEME 2.17   Mechanism of synthesis of pyrazolo-oxazin-2-thione compounds.

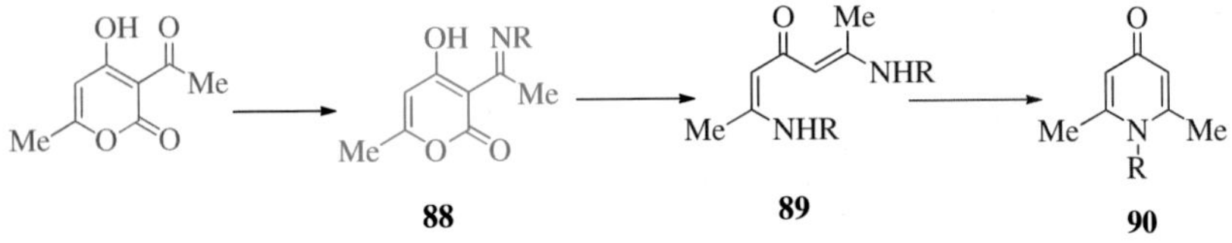

SCHEME 2.18   Synthesis of pyridones.

gives 2,6-dimethyl-4-pyridones (**90**) followed by the formation of intermediates [82–86] (**88, 89**) (Scheme 2.18).

Clopidol [87] (**2**) (chloro derivative of 2,6-dimethyl-4-pyridone) is synthesized from 2,6-dimethyl-4-pyridone (**90**). It can be used as a coccidiostatic agent. Amprolium (**91**) (INN, trade name Amprol) is one of the coccidiostatic agents used in poultry

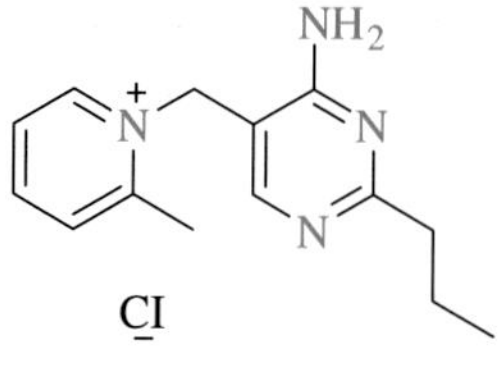

Interesting versions of reactions [88–90] arise with hydrazine and N-amino heterocyclic compounds. The reactions of dehydroacetic acid with N-aminopyridinium salts, N-amino triazole, and hydrazine yields various compounds (**92, 93,** and **94**) (Scheme 2.19).

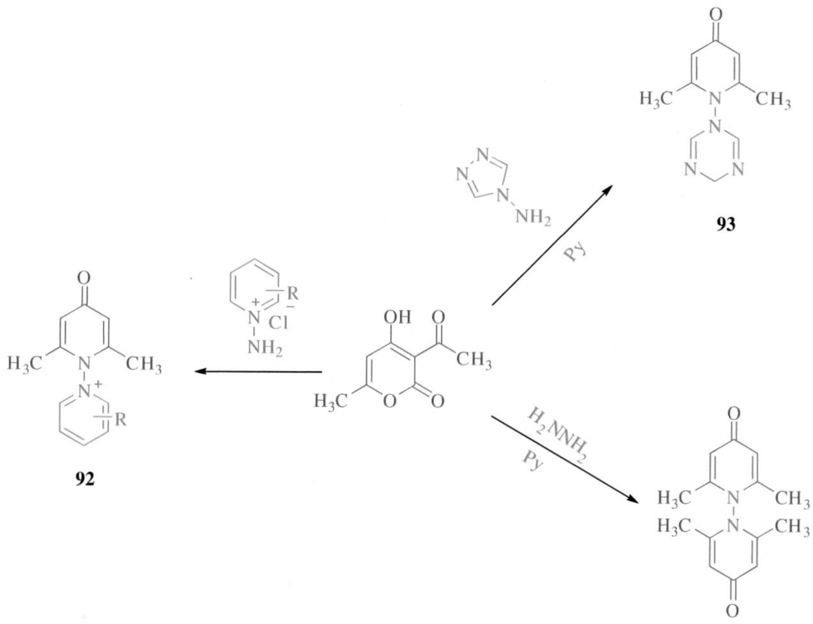

SCHEME 2.19 Synthesis of triazinyl pyridinones and bipyridine-4'-diones.

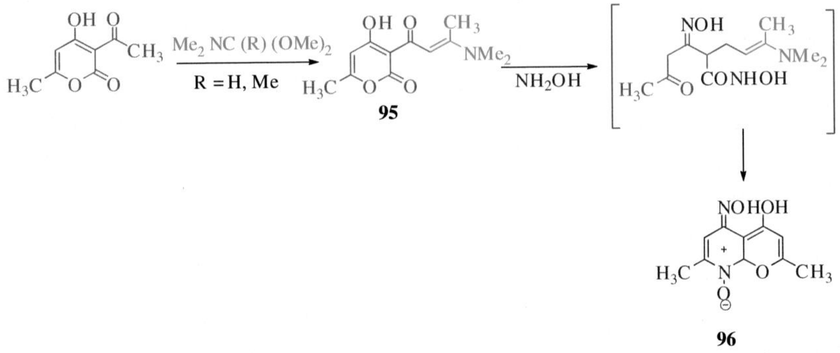

SCHEME 2.20 Synthesis of pyrano pyridine olates.

2.6 SYNTHESIS OF PYRANO PYRIDINEOLATES

DHA on reaction [91] with two moles of hydroxylamine forms products (**95**) and (**96**). The final products (**95** and **96**) formed are the best examples of the complexities encountered in opening cyclization sequences (Scheme 2.20).

2.7 SYNTHESIS OF PYRIDINOL AND DIMETHYL NICOTINIC ACID

In 1885, the famous scientist named Haitinger was the first to report the formation of dimethyl pyridinol (**97**) and hydroxydimethylnicotinic acid (**98**) by the action of **DHA** (**1**) with aqueous ammonia [83]. Since then, much research has been undertaken to define the mechanism of the reaction taking place between **DHA** and alkyl amines. When the reaction takes place between **DHA** and **NH₃**, six by-products will be formed. The mechanistic clarification and methods to decrease the generation of unwanted by-products has been explored by many researchers, achieving 97% yield of compound (**97**). The remaining six other by-products are mentioned in Scheme 2.21.

The mechanisms involved in the formation of compounds (**99–103**), which achieve less than 15% yield are not described. In the **DHA** structure, four possible sites are very important. They are the lactone carbonyl group (>C=O) at second position, the carbonyl group (>C=O) of the acetyl side chain at third position, the hydroxyl group attached to the carbon atom at fourth position, and the carbon atom of the methyl group attached at sixth position, which could be attacked by ammonia.

The attack of ammonia on the second position (lactocarbonyl group) of **DHA** affords acetylhydroxymethylpyridone (**99**). Then it is hydrolyzed by hydroxides of alkali metals (caustic alkali) to produce the compound (**100**) (Scheme 2.22).

When ammonia binds to the carbonyl group of the acetyl side chain which is present at the third position of **DHA** (**1**), a Schiff base named imine (**104**) (crystalline compound) is formed with more than 90% yield at room temperature. When the compound (**104**) is heated with aqueous NH₃ or H₂O at 200°C,

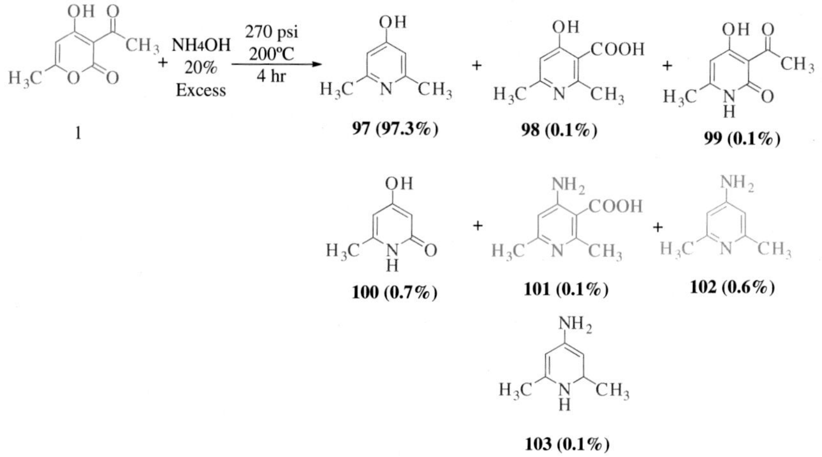

SCHEME 2.21   Synthesis of pyridinol and dimethyl nicotinic acid.

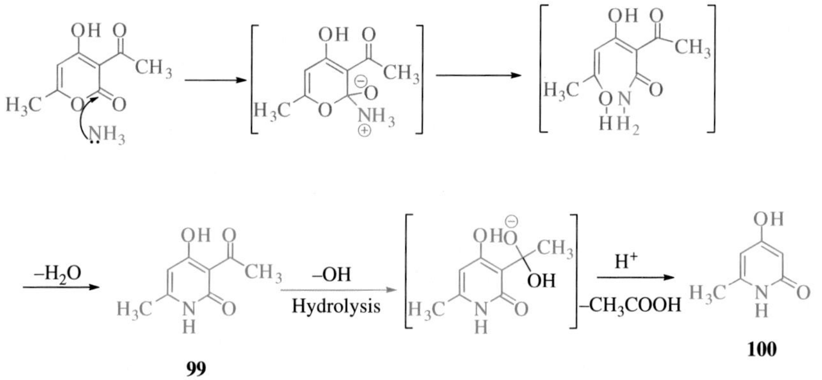

SCHEME 2.22   Synthesis of hydroxy-methylpyridin-2-ones.

compounds (**97–103**) will be formed. The conversion of compound (**104**) into 2,6-dimethyl-m-pyridinol (**97**) and hydroxydimethylnicotinic acid (**98**) is shown in Scheme 2.23.

The attack of **NH₃** at the sixth position of **DHA** can produce the compounds (**97 and 98**). However, this scheme became less important due to the excessive yields of compound (**104**) at normal room temperature.

Ammonia attacks on the fourth position of **DHA** (**1**) form the product acetylaminomethylpyranone (**105**). In the next step, if the compound (**105**) follows **path a**, it forms aminolutidine (**102**), amino-dimethyl nicotinic acid (**101**). Finally, if it (**compound 105**) undergoes **path b**, it yields 4-amino-6-methyl-pyridin-2(1*H*)-one (**103**) (Scheme 2.24).

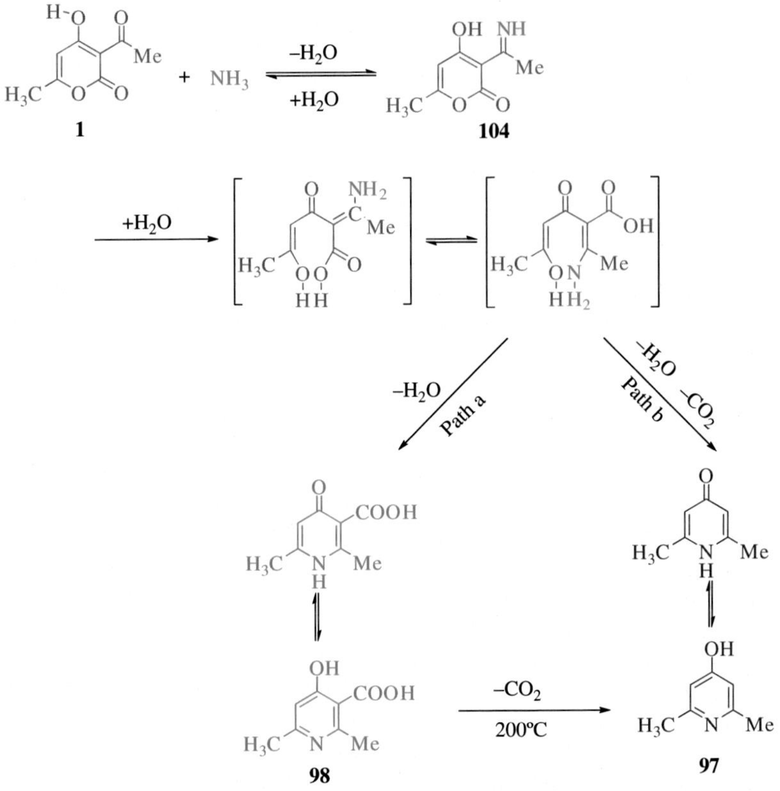

SCHEME 2.23   Synthesis of dimethyl-m-pyridinol and dimethyl nicotinic acid.

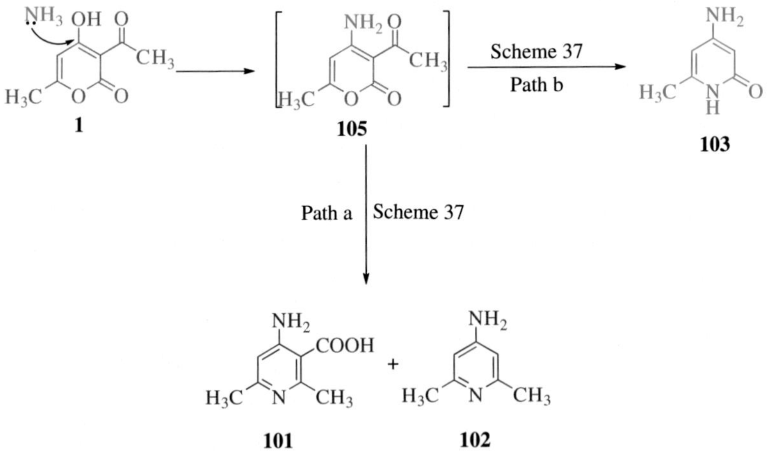

SCHEME 2.24   Synthesis of amino lutidine, amino-dimethyl nicotinic acid and amino-methyl pyridinone.

When dehydroacetic acid (**1**) reacts with 20% excess amount of ammonia at 200°C for 1 hour, it will be converted into dimethyl pyridinol (**97**) and hydroxy-dimethylnicotinic acid (**98**) along with limited amounts of compounds (**99, 100, 101, 102,** and **103**). During the decarboxylation of hydroxy-dimethyl nicotinic acid (**98**) to dimethylpyridinol (**97**) at 200°C, the compounds (**101, 102,** and **103**) will be almost completely formed. Then the compounds (**97, 98,** and **100**) react with the residual ammonia solution to produce aminolutidine (**102**), amino-dimethyl nicotinic acid (**101**), and aminomethylpyridone (**103**), respectively. It must be observed that 50% of compounds (**97**) or (**100**) will get converted to (**102**) or (**103**) at 250°C within 6 hours.

2.8 SYNTHESIS OF VARIOUS PYRIMIDINES

DHA reacts with thiourea [92] in the presence of piperidine to afford 1-(6-methyl-2-thioxo-2,5-dihydropyrimidin-4-yl)propan-2-one (**106**). With amino thiourea [93], it produces 2,5-dimethyl-7-thioxo-6,7-dihydropyrazolo(1,5-C) pyrimidine (**107**) which has insecticidal and fungicidal activity. With guanidine, it is transformed to 1-(2-amino-6-methylpyrimidin-4-yl)propane-2-one (**108**), and with phenyl guanidine nitrate [94] and on heating with potassium carbonate in DMF, it forms anilino pyrimidine (**109**), its chloro derivative is a known fungicide (Scheme 2.25).

2.9 REACTIONS THAT MODIFY THE 2-PYRONE SKELETON

DHA undergoes ring-opening reactions in the presence of appropriate reagents and produces different cyclic products without any transformations. These types of reactions are comparatively uncommon, although some of them are

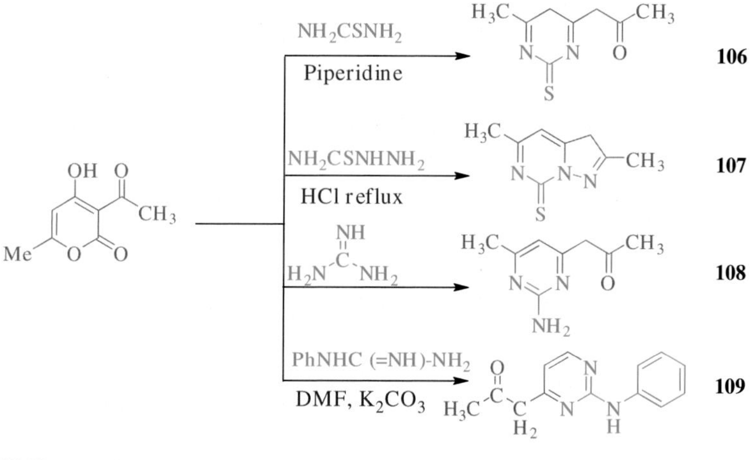

SCHEME 2.25 Synthesis of various pyrimidines.

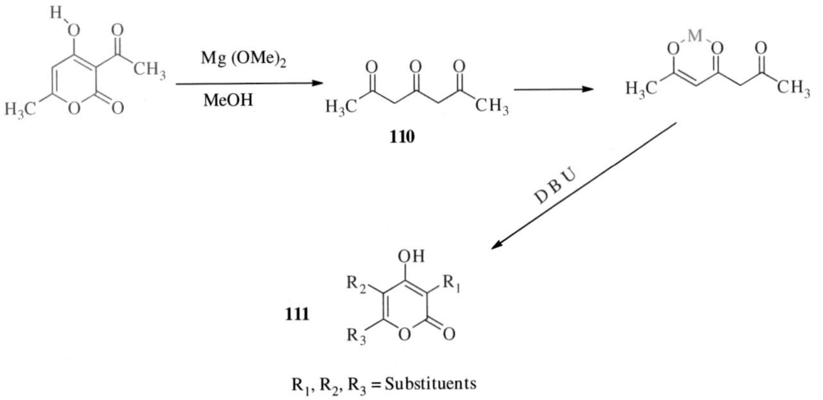

$R_1, R_2, R_3 =$ Substituents

SCHEME 2.26 Synthesis of substituted pyrones by DHA ring opening.

very important. **DHA** can effectively convert to methyl-3,5-dioaxahexanoate [95] (**110**) by reaction with magnesium methoxide in the presence of methanol.

2.9.1 Synthesis of Various Substituted Pyrone by DHA Ring-Opening

By using diazabicycloundecane (**DBU**) in benzene, the diketoester (**110**) (polyketide model) can be cyclized and regioselectively alkylated at C_3 and C_5 which provides a route to a large array of pyrenes [96–99] (**111**) (Scheme 2.26).

2.9.2 Synthesis of Diones

DHA readily reacts with pyrrolidine to afford amino heptane dione (**112**) and finally it forms dienone (**113**) by ring-opening and decarboxylation (Scheme 2.27).

2.9.3 Synthesis of 2,4,6-Triones

DHA undergoes hydrolysis and decarboxylation [100] to form heptane-2,4,6-trione (**110**) (Scheme 2.28).

2.10 SYNTHESIS OF BENZODIAZEPINE, BENZODIAZEPINONE, AND BEZIMIDAZOLES

The reaction [101] of o-phenylenediamine with **DHA** in the presence of different alcohols forms 2-(alkoxy carbonyl methylene) benzodiazepine (**114**), 4-(acetyl methylene) benzodiazepinone (**115**), and benzimidazoles (**116** and **117**) (Scheme 2.29).

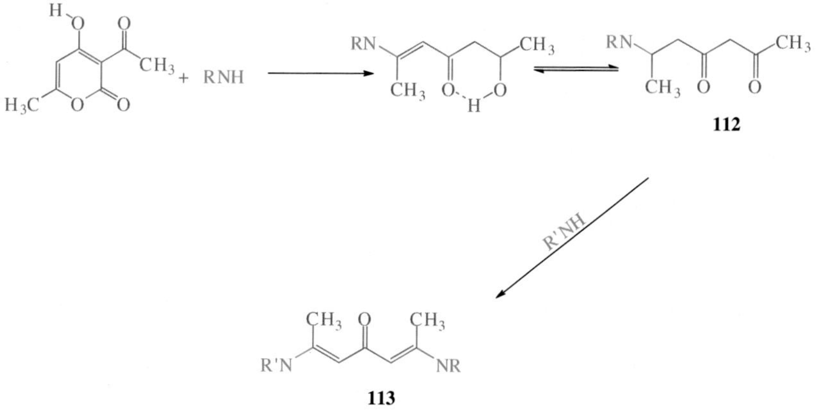

SCHEME 2.27   Synthesis of diones.

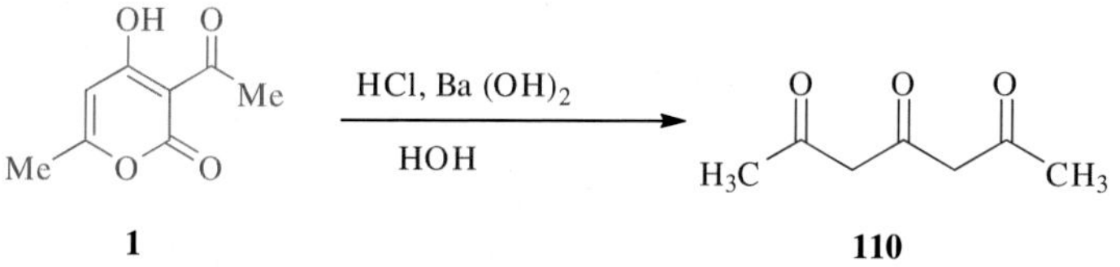

SCHEME 2.28   Synthesis of 2, 4, 6 –triones.

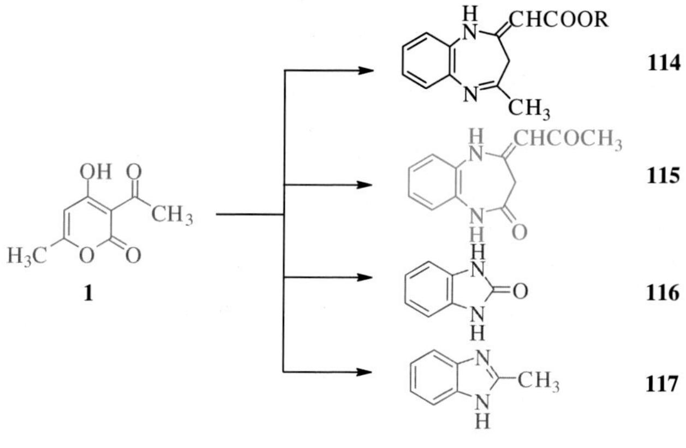

R = Me, Et, Pr, Bu

SCHEME 2.29   Synthesis of benzodiazepine, benzodiazepinone and bezimidazoles.

Chapter 3

Synthesis of Different Heterocyclic Compounds by Using DHA

Gudala Satish

National Institute of Technology Raipur, Raipur, Chhattisgarh, India

3.1 SYNTHESIS OF PYRAZOLES FROM VARIOUS CHALCONES

Dehydroacetic acid reacts with different types of aldehydes to produce different types of chalcones. These chalcones are very important intermediates to synthesize a large quantity of biologically active dehydroacetic acid derivatives. Here, we will discuss the synthesis of chalcones and pyrazoles from chalcone intermediates [102]. The chalcone (**118**) was synthesized from **DHA** and oxaldehyde in the presence of piperidine using chloroform as a solvent. To the solution of chalcones dissolved in glacial acetic acid, phenyl hydrazine was added and the mixture was refluxed for 4 hours. It was then cooled and neutralized with NH_4OH to produce pyrazoles (**119**) (Scheme 3.1).

DHA reacted with terephthalaldehyde in the presence of piperidine using chloroform as a solvent, producing DHA substituted chalcone (**120**). This chalcone was further reacted with phenyl hydrazine and hydrazine to give various pyrazoles (**121 and 122**) (Scheme 3.2).

DHA reacted with 2-bromobenzaldehyde in the presence of piperidine using chloroform as a solvent, producing good chalcones, namely (*E*)-3-(3-(2-bromophenyl)acryloyl)-4-hydroxy-6-methyl-2*H*-pyran-2-one (**123**). This chalcone was further reacted with phenyl hydrazine and *p*-nitrophenyl hydrazine to give various pyrazoles (**124 and 125**) (Scheme 3.3).

DHA reacted with 2,5-dimethoxybenzaldehyde in the presence of piperidine using chloroform as a solvent, producing good chalcones, namely (*E*)-3-(3-(2,5-dimethoxyphenyl)acryloyl)-4-hydroxy-6-methyl-2*H*-pyran-2-one (**126**) (Scheme 3.4). This chalcone was further reacted with hydrazine in the presence of formic acid to give various pyrazoles (**127**). In another path, the chalcone reacted with hydrazine in the presence of acetic acid to give acetyl

Dehydroacetic Acid and Its Derivatives. DOI: http://dx.doi.org/10.1016/B978-0-08-101926-9.00003-1
© 2017 Elsevier Ltd. All rights reserved.

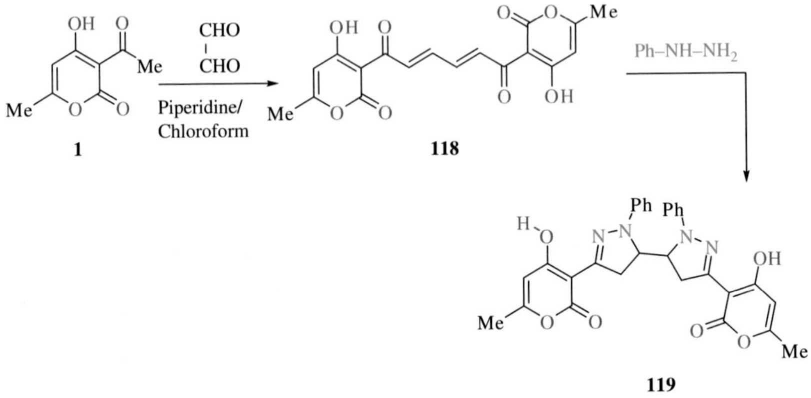

SCHEME 3.1 Synthesis of Pyrazoles from various chalcones.

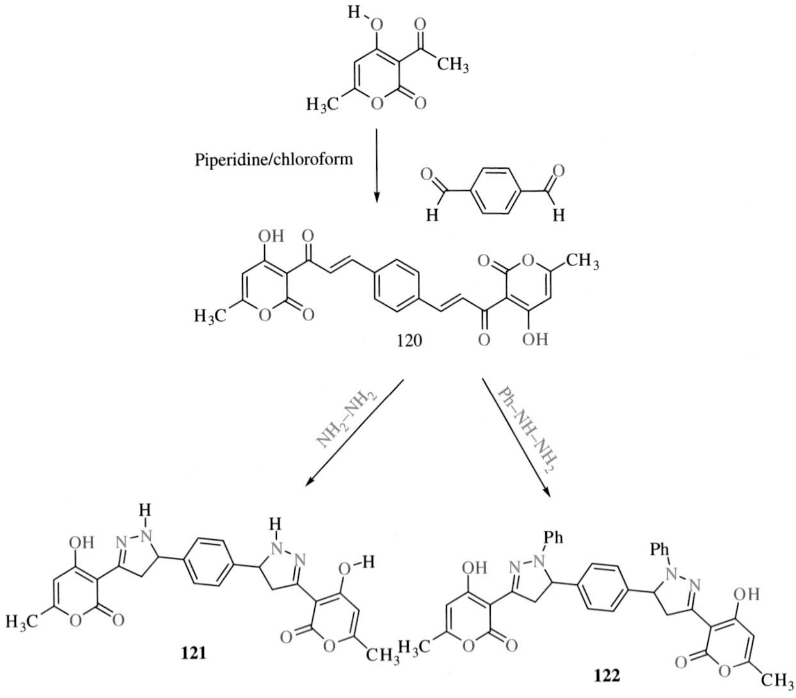

SCHEME 3.2 Synthesis of pyrazoles from chalcones.

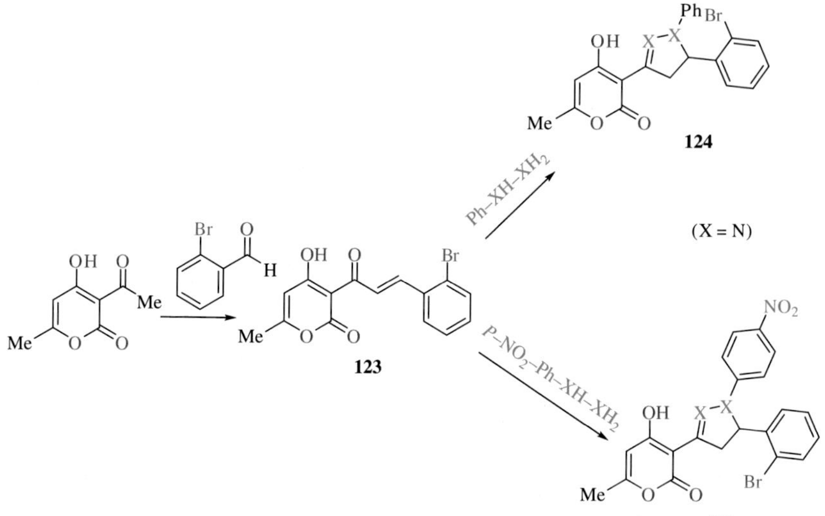

SCHEME 3.3 Synthesis of pyrazoles from phenyl hydrazine and p-nitrophenyl hydrazine.

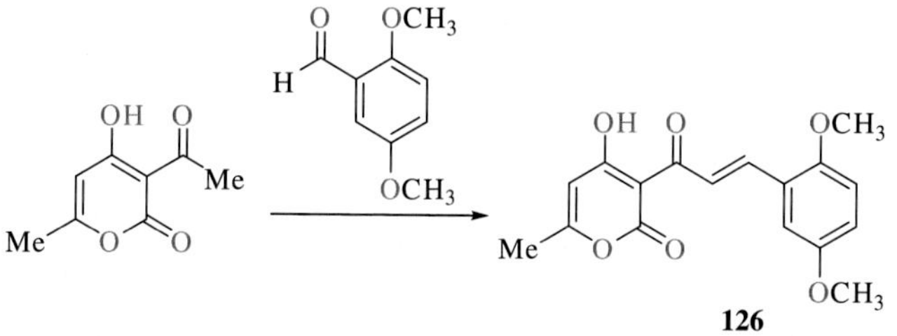

SCHEME 3.4 Synthesis of dimethoxyphenyl-acryloyl-hydroxymethyl-2H-pyran-2-ones.

pyrazoles. These acetyl pyrazoles again reacted with 2-bromobenzaldehyde, producing (*E*)-3-(1-(3-(3-bromophenyl)acryloyl)-5-(2,5-dimethoxyphenyl)-4,5-dihydro-1*H*-pyrazol-3-yl)-4-hydroxy-6-methyl-2*H*-pyran-2-one (pyrazoles **128**) (Scheme 3.5).

Chalcone (**126**), which was previously synthesized, was again treated with 2-hydrazinylbenzo[d]oxazole and 2-hydrazinyl-1*H*-benzo[d]imidazole to give substituted pyrazoles (**129** and **130**) (Scheme 3.6).

The compound (**131**) was synthesized from **DHA** by its reactions with *p*-amino ethyl benzoate in the presence of sodium ethoxide. Then, this inter-mediate (**131**) was again treated with hydrazine and produced the pyrazoles 3-(5-(4-aminophenyl)-4,5-dihydro-1*H*-pyrazol-3-yl)-4-hydroxy-6-methyl-2*H*-pyran-2-one (**132**) (Scheme 3.7).

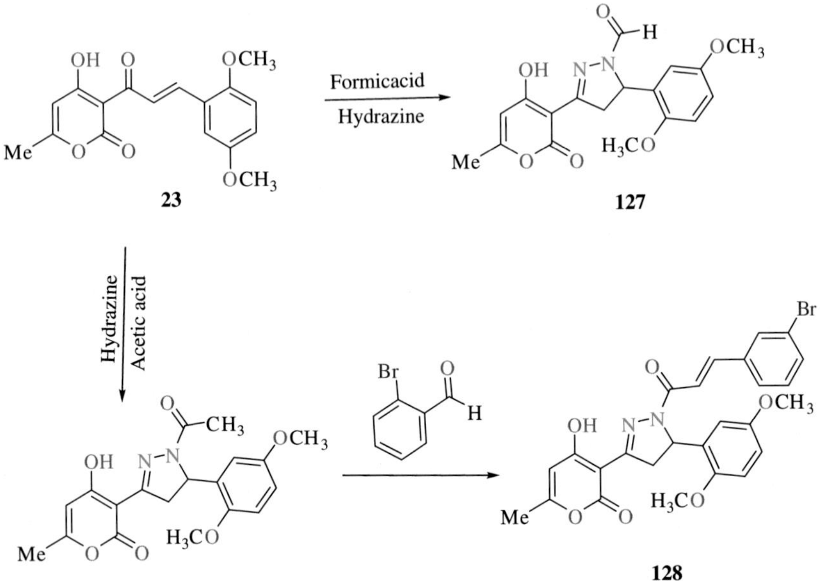

SCHEME 3.5   Synthesis of bromophenyl acryloyl-dimethoxyphenyl-dihydropyrazolyl-hydroxy-methyl-2H-pyran-2-one.

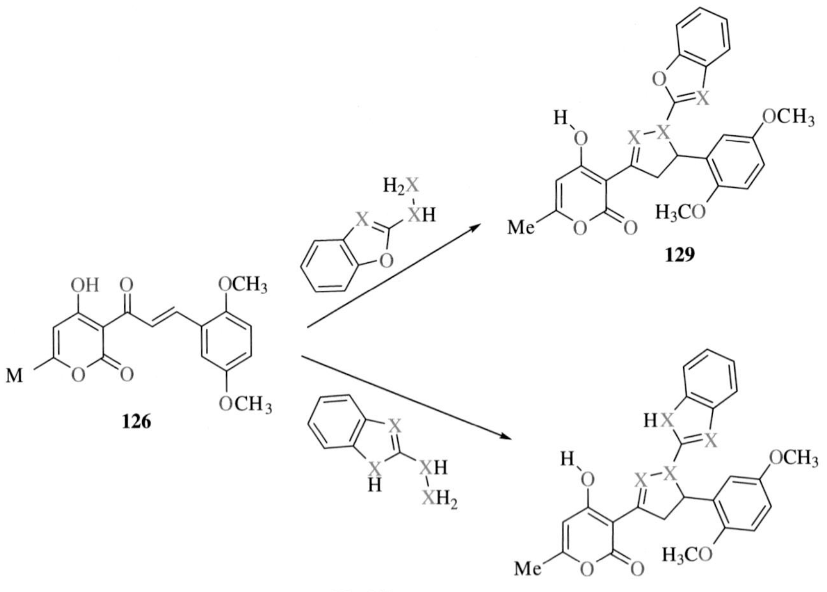

X = Nitrogen

SCHEME 3.6   Synthesis of pyrazoles from hydrazinyl-benzooxazole and hydrazinyl-benzoimidazole.

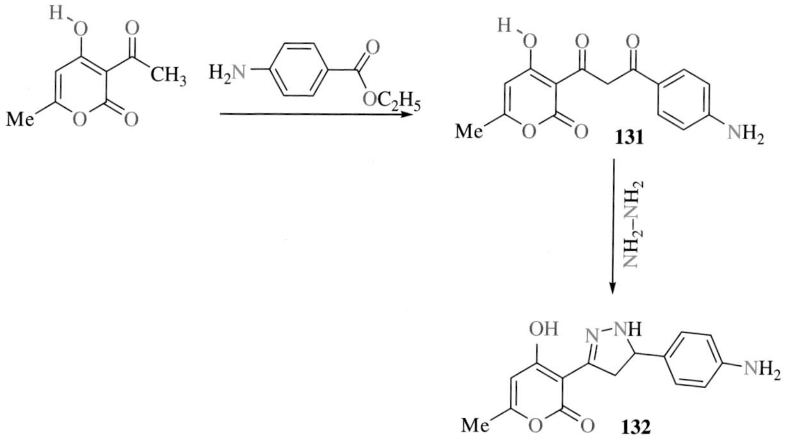

SCHEME 3.7   Synthesis of pyrazoles from p-amino ethyl benzoate and hydrazine.

3.2 SYNTHESIS OF SUBSTITUTED 2*H*-PYRAN-2-ONES

The Knoevenagel condensation of **DHA** with aromatic aldehyde in the presence of pyridine and piperidine as catalyst under thermal refluxing or microwave irradiation in CHCl$_3$ as a solvent leads to the synthesis of compounds (**133a–c** and **135**).

The selective hydrogenation reaction takes place in arylpropenoyl substituted **DHA** (**133a–c**) and phenylpentadienoyl substituted **DHA** [103] (**135**) in the presence of catalyst (10% Pd/C) and under the pressure of H/CH$_3$COOC$_2$H$_5$ for the synthesis of arylpropenoylpyranones (**134a–c**) and phenylpentanoylpyranone (**136**) (Scheme 3.8).

The synthesized compounds arylpropenoylpyranones (**134a–c**) and phenylpentanoylpyranone (**136**) reacted with *o*-phenylenediamine (**137**) in the presence of C$_2$H$_5$OH at room temperature for 6 hours to give the best yields (**Method A**). But in **Method B** it should reflux under thermal conditions for about 30 minutes. **Method C** uses microwave conditions at 100°C for 1 minute and for **Method D** it should be kept for 4 minutes. Methods **A**, **B**, **C**, and **D** separate the final compounds **138a–c** and **139**, whereas **Method D** individually separates the compound, namely the imidazole derivative of **DHA** (**140**) (Scheme 3.9).

Method A: 6 hours at normal room temperature; **Method B:** 1 hour thermal refluxing; **Method C:** 1 minute under microwave irradiation at 100 W; **Method D:** 4-minute microwave irradiation at 100 W.

The derivatives of the above compounds are listed in Table 3.1.

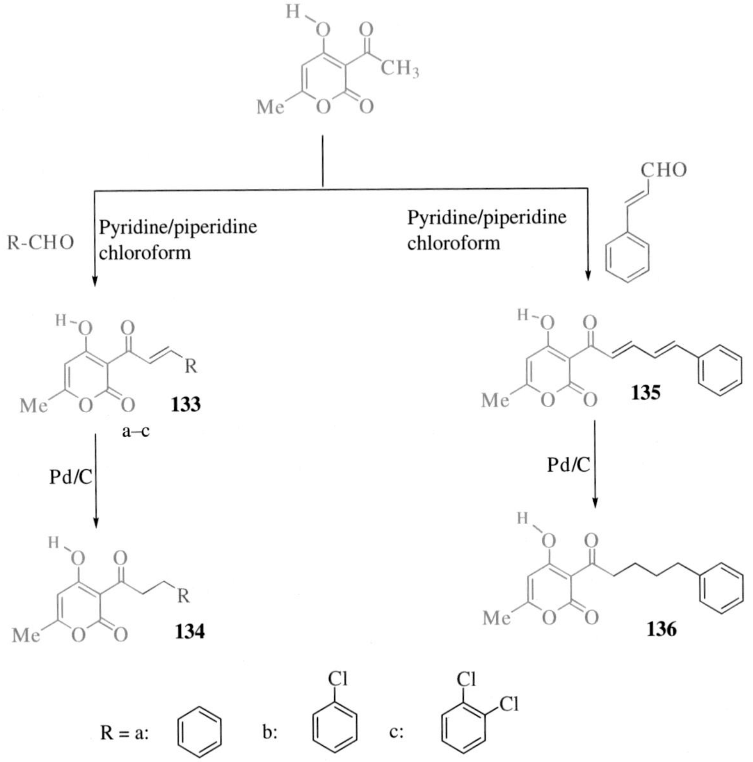

SCHEME 3.8 Synthesis of substituted 2H-pyran-2-ones.

3.3 SYNTHESIS OF VARIOUS *P*-TOLYLACRYLOYL DERIVATIVES OF DHA

Dehydroacetic acid with aromatic aldehydes (**141**) using microwave-assisted Knoevenagel condensation leads to the formation of various *p*-tolylacryloyl derivatives of **DHA** [104] and its derivatives (**142a–h**) (Scheme 3.10). When compared with the conventional method, the reaction was executed under solvent-free conditions in MW irradiation to improve the rate of reaction.

Derivatives of **142a–h** are listed in Table 3.2.

For the synthetic procedures for various *p*-tolylacryloyl derivatives of DHA, a mixture of **DHA** and aromatic aldehydes are taken in a vessel containing neutral alumina. To this, a pyridine and piperidine mixture is added as a catalyst and the reaction mixture is irradiated for about 4–10 minutes under microwaves at 200 W power. Then, the product is extracted using chloroform and the crude/final product is recrystallized from ethanol.

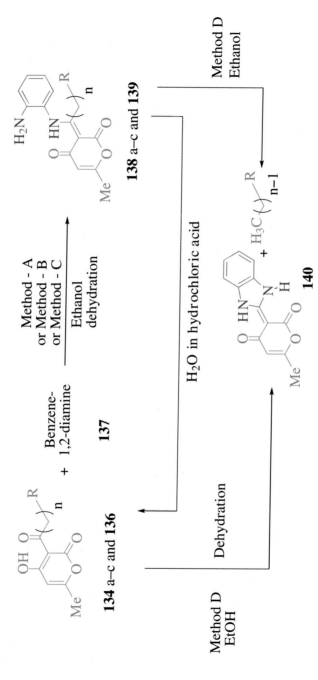

SCHEME 3.9 Synthesis of imidazoles and phenylhydrazine derivatives of DHA.

TABLE 3.1 Derivatives of Substituted 2*H*-Pyran-2-one

134a	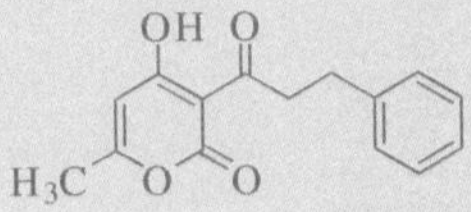	**138a**	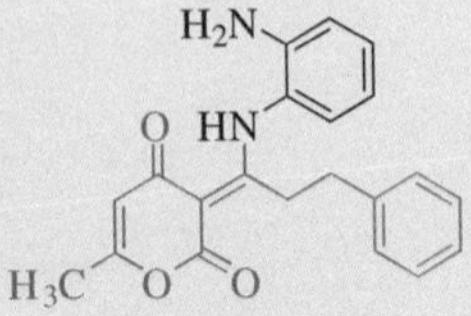
134b	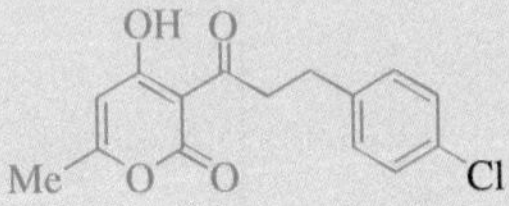	**138b**	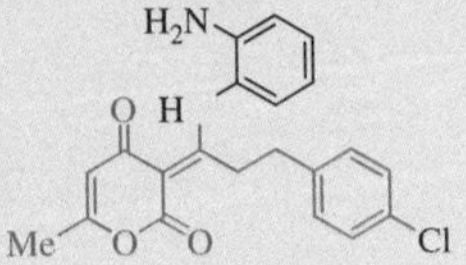
134c	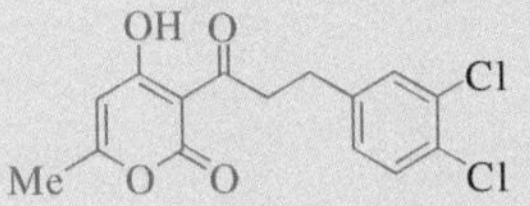	**138c**	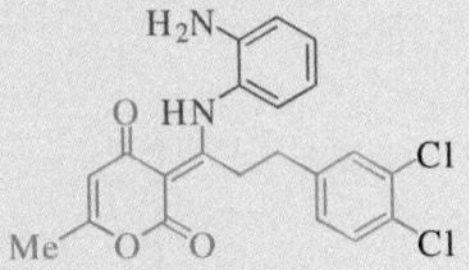
136	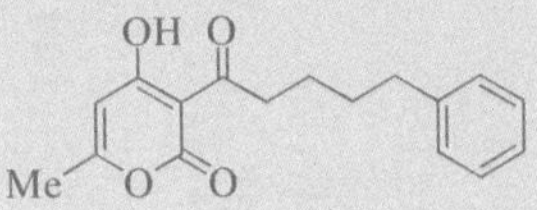	**139**	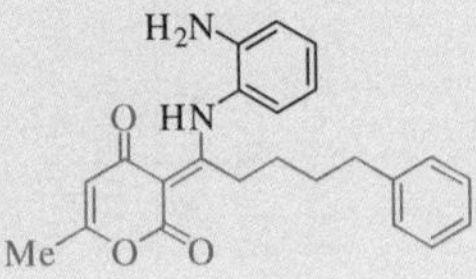

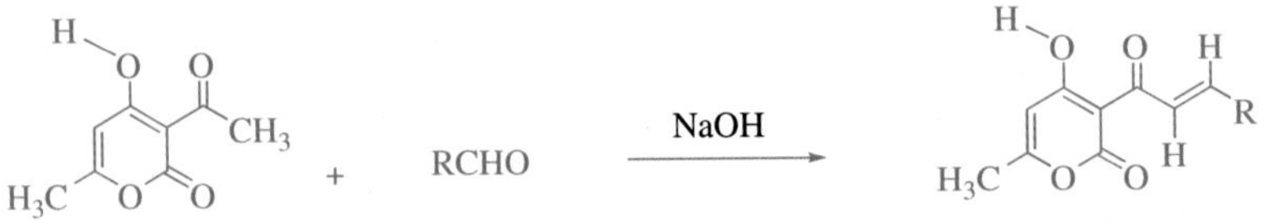

1 **141** **142** a–h

SCHEME 3.10 Synthesis of various p-tolylacryloyl derivatives of DHA.

3.4 CLASSICAL PROCEDURE FOR SYNTHESIS OF PHENYLIMINO DERIVATIVES OF DHA

Phenylimino derivative of **DHA** (**143**) was synthesized by charging the 100 mL round bottomed flask with a mixture of **DHA**, amine, and ***p*-TSA** in ethanol in a ratio of 4:4:2. Next, these compounds were stirred and refluxed at 37°C. Then, the final compound was cooled and under reduced pressure, the other solvents were evaporated. Finally, the compound was washed with EtOH to get the pure solids.

TABLE 3.2 The *p*-tolylacryloyl derivatives of DHA 142a-h

S.NO.	DERIVATIVES	S.NO.	DERIVATIVES
142a	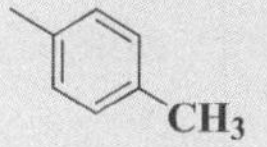 CH₃	142e	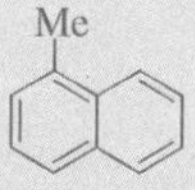 Me
142b	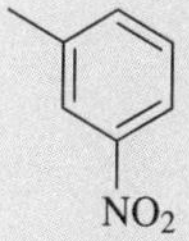 NO₂	142f	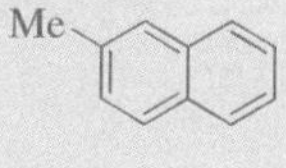 Me
142c	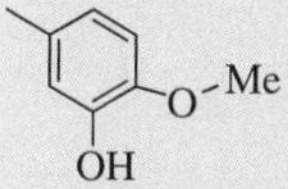 O–Me OH	142g	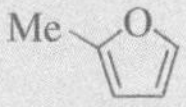 Me O
142d	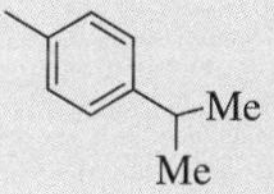 Me Me	142h	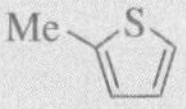 Me S

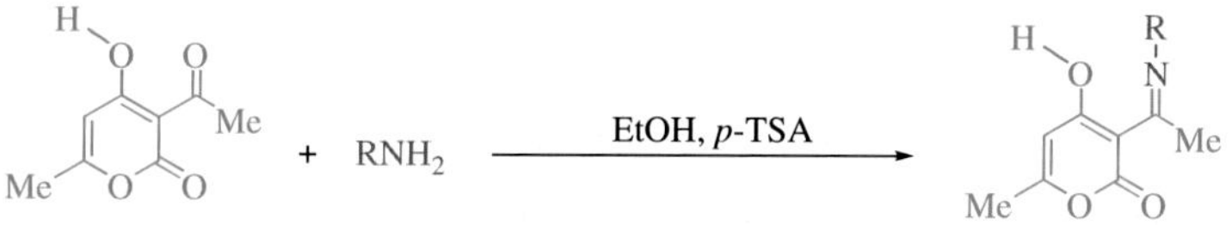

SCHEME 3.11   Synthesis of phenylimino derivatives of DHA.

3.4.1   Synthesis of Phenylimino Derivative of DHA Derivatives by Using the Ultrasound Promoted Method

Phenylimino derivative of **DHA (143)** are also synthesized by using the ultrasound promoted method. In this method, a mixture of **DHA**, amine, and *p*-TSA in ethanol in a ratio of 2:2:1 was charged in the 100 mL round bottomed flask, and the mixture was placed in a sonicator at 30°C. Then the product was cooled and under reduced pressure, the other solvents were evaporated. Finally, the compound was washed with EtOH to get the pure solids [105] (Scheme 3.11).

3.5   SYNTHESIS OF DHA SUBSTITUTED THIAZOLES

DHA undergoes bromination with bromine in CHCl₃ to give a new product, tribromo **DHA (144)**. The compound (**144**) reacts with phenyl thiourea (**145**) in ethanol at room temperature to give a solid compound bromodehydroacetic

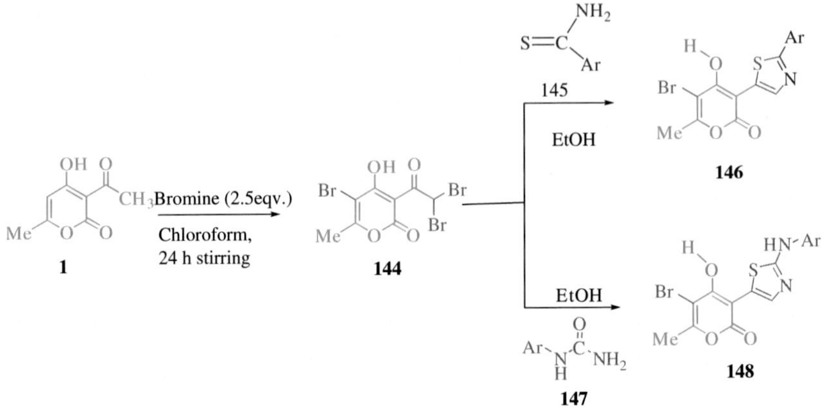

SCHEME 3.12   Synthesis of thiazoles.

acid–aminophenylthiazole (**146**) which will be separated within 10 minutes of the reaction starting time [106]. The compound (**144**) reacts with a variety of thiourea in a similar manner to form different types of substituted thiazoles (Scheme 3.12). In the above reaction, the thiourea can be replaced by thiomides (**147**) to afford arylbromo hydroxyl methylpyranylthiazoles (**148**). The reaction has large advantages, such as mild reaction conditions, short reaction time, high yield, convenient method for the separation of final compounds and solids without any purifications, and also the intact nature of the pyranyl moiety of **DHA**.

3.6   SYNTHESIS OF DIAZEPINES

Diazepines (**153**) can be synthesized in two steps [107] by using three versatile compounds namely **DHA**, 1,2-diamine, and Ar-CHO. Initially **DHA** reacts with 1,2-diamine (**149**) to afford intermediates, namely ketimines (**Compound A (150)** or **Compound B (151)**) (Scheme 3.13).

So, on the basis of the nucleophilic nature of nitrogen N_4 and N_5, the Compound A (through nitrogen N_5) or Compound B (through nitrogen N_4) will be formed (Scheme 3.14).

In the next step, the final 1,4-diazepine derivative compound will be formed by the reaction between ketimine (**Compound A**) and benzaldehyde (Scheme 3.15).

The reaction mechanism is given in Scheme 3.16.

In the mechanism, the benzaldehyde reacted with the ketimine intermediate (**Compound A**) and formed the benzylideneaminomethylpropylimino derivative of **DHA** by losing a water molecule which is again followed by two paths to yield the diazepine compound (**153**) as a final product.

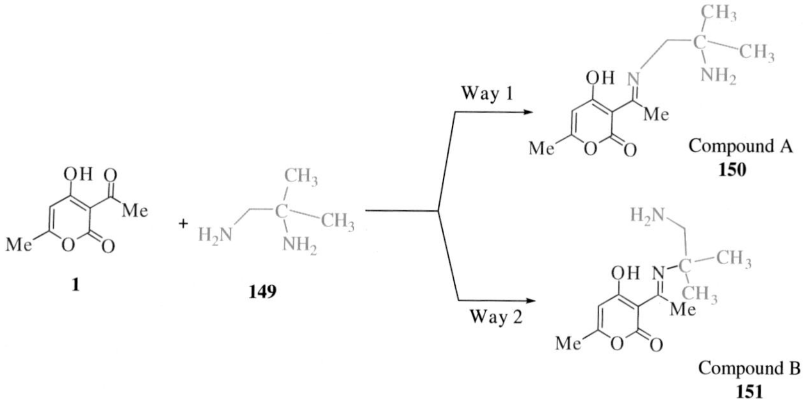

SCHEME 3.13 Synthesis of diazepines.

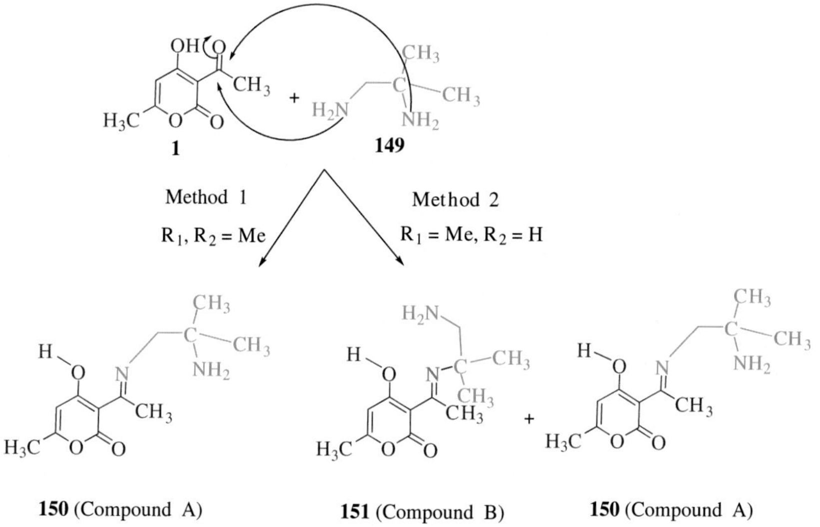

SCHEME 3.14 Formation of ketimines from DHA and 1, 2-diamine.

3.6.1 Diazepines and Benzodiazepines

DHA on reaction with 2-methylpropane-1,2-diamine in the presence of ethanol afforded the compound (Z)-3-(1-(2-amino-2-methylpropylimino)ethyl)-4-hydroxy-6-methyl-2H-pyran-2-one (Enaminones). It was further reacted with different aldehydes in the presence of ethanol (Keggin type HPA catalysts [108]) to form 1,4-diazepines (**154**). On the another hand, **DHA** reacted with benzene-1,2-diamine in the presence of ethanol producing one of the enaminones,

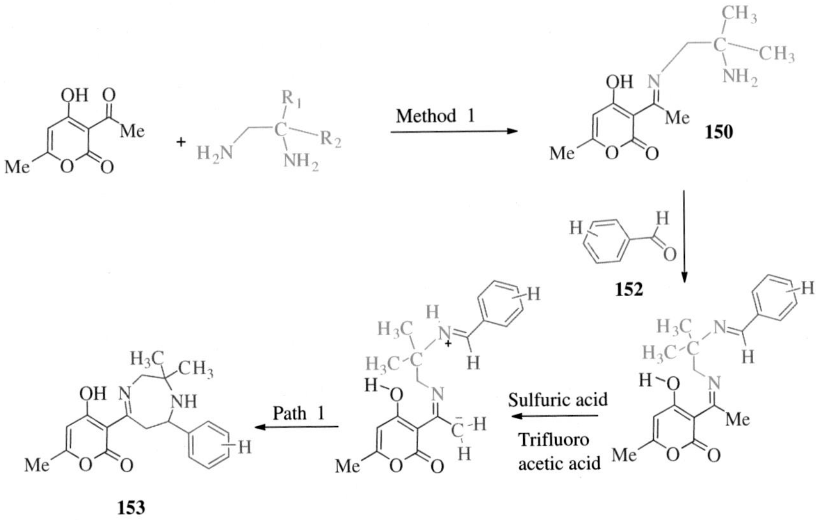

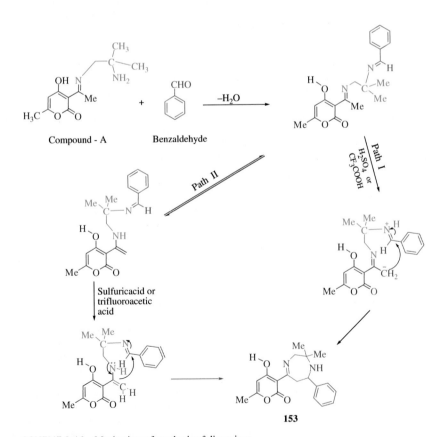

SCHEME 3.15 Formation of 1, 4-diazepines from ketimine and benzaldehyde.

SCHEME 3.16 Mechanism of synthesis of diazepines.

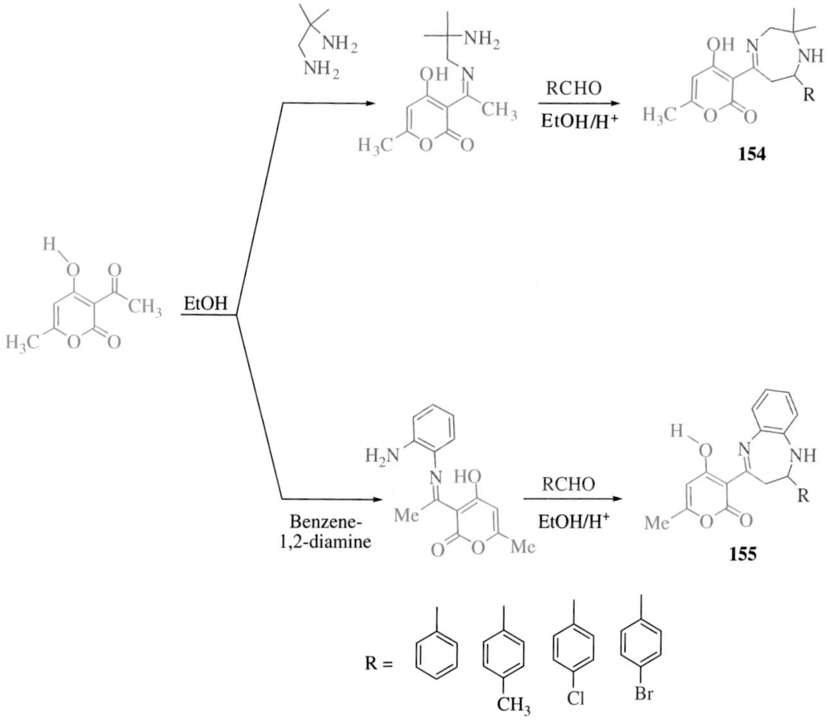

SCHEME 3.17 Synthesis of 1,4diazepines and 1,5 benzodiazepines.

namely the aminophenylimino derivative of **DHA**. Then, it was further reacted with different aldehydes in the presence of ethanol and produced 1,5-benzodiazepines (**155**) (Scheme 3.17).

Derivatives of the compounds **154** and **155** are listed in Table 3.3.

The reaction mechanism is given in Scheme 3.18.

3.6.2 Benzodiazepines Synthesis by Using Bismuth (III) Catalysis

The reaction of the ketimine derivative of **DHA** in MeOH (**156**) with DMF (**157**) in the presence of bismuth triflate or $BiCl_2$ gave the final compound 4-pyrano-1,5-benzodiazepines. Then, the mixture was refluxed under magnetic stirring for 48 hours, allowed to cool, and filtered to get the final product, namely benzodiazepine (**159**) [109]. The reaction mechanism is given in Scheme 3.19.

The benzodiazepines (**159**) synthesis can be schematized into two steps. In the first step, the formation of the diimine intermediate (**158**) takes place by the interaction between the amino group of aminophenylimino derivatives of **DHA** (**156**) and the masked aldehyde (**157**) carbonyl group by eliminating two methyl

TABLE 3.3 Derivatives of 1,4diazepines 154 and 1,5 benzodiazepines 155

Compounds	R
154a	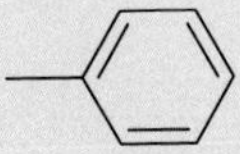
154b	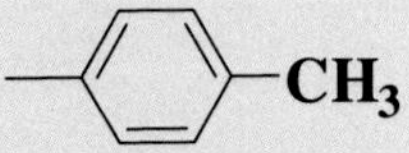 —⟨phenyl⟩—CH₃
154c	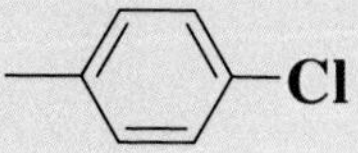 —⟨phenyl⟩—Cl
154d	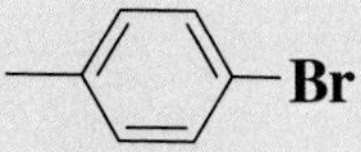 —⟨phenyl⟩—Br
155a	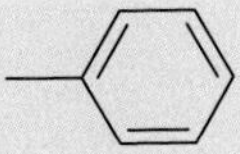
155b	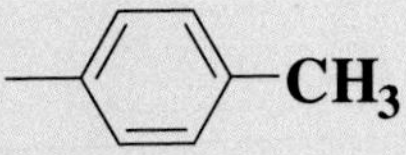 —⟨phenyl⟩—CH₃
155c	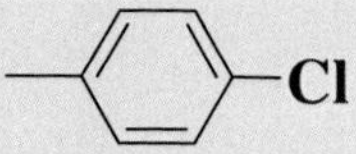 —⟨phenyl⟩—Cl
155d	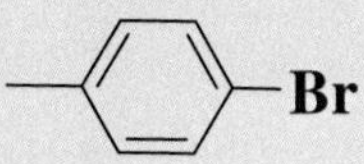 —⟨phenyl⟩—Br

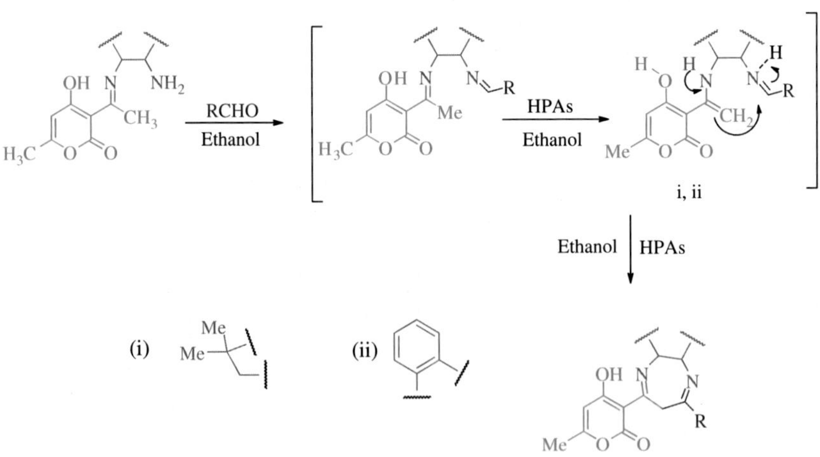

SCHEME 3.18   Mechanism of synthesis of 1,4diazepines and 1,5 benzodiazepines.

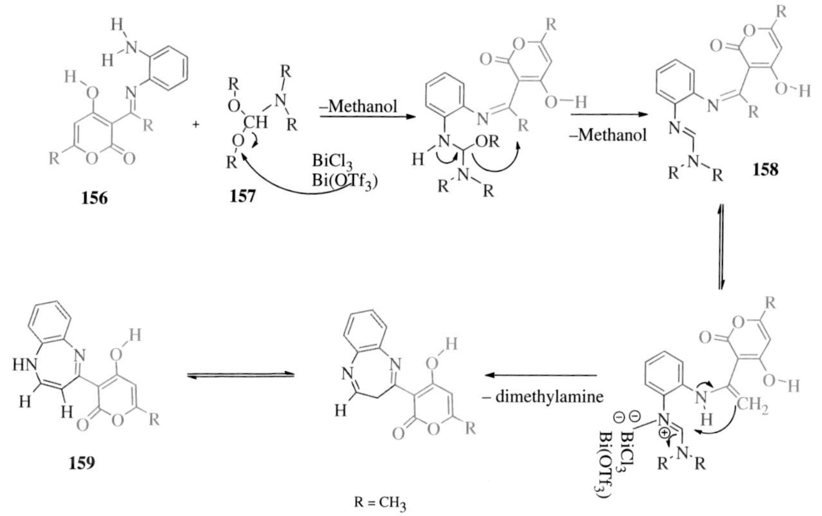

SCHEME 3.19 Synthesis of benzodiazepines by using bismuth (III) catalysis.

alcohol molecules. In the second step, the diimine intermediate (158) undergoes cyclization under acidic catalysis which leads to the synthesis of benzodiazepine (159) by the protonation of the imino group and methyl group mobility because of the hyperconjugation. Actually, the benzodiazepine (159) derivative exists in a tautomeric form, which was supported by its spectral data.

3.6.3 Benzothiazepines and Benzothiazines

3.6.3.1 1,5-Benzothiazepine (162) and 1,4-Benzothiazine (163)

The treatment of α,β-unsaturated ketones (161) with o-aminothiophenol (o-ATP) (160) afforded 1,5-benzothiazepines [110]. This is one of the most important and widely used methods for the preparation of 1,5-benzothiazepines (162) in the presence of acidic and basic conditions (Scheme 3.20).

As in the above reaction, the treatment of α,β-unsaturated ketones (161) o-ATP (160) followed by the 6-exo-mechanism produced dihydro-1,4-benzothiazines (163) (Scheme 3.21).

Hence, according to Baldwin's rules, the **7**-endo-trig procedure leads to the formation of 1,5-benzothiazepine (162), whereas the 6-exo-trig procedure leads to the formation of dihydro-1,4-benzothiazine (163). Both these compounds (162 and 163) are isomers and cannot be differentiated easily.

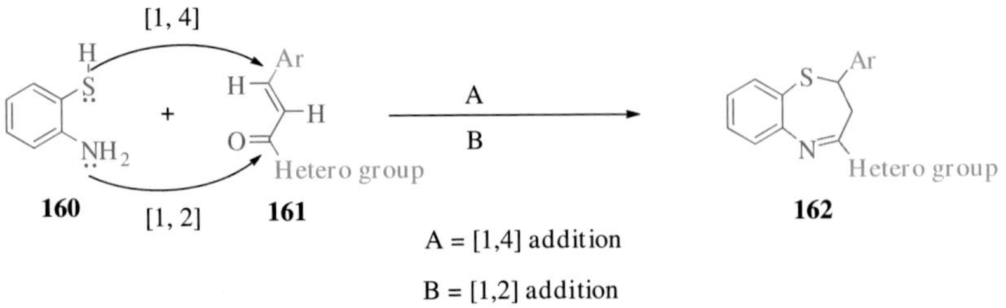

SCHEME 3.20   Synthesis of 1, 5-benzothiazepines.

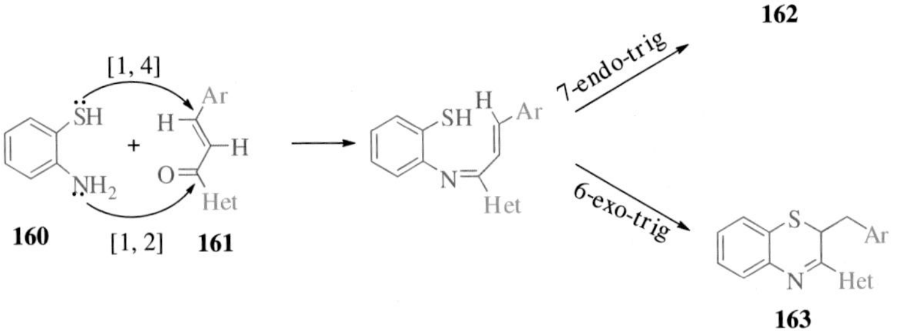

SCHEME 3.21   Synthesis of dihydro-1, 4-benzothiazines.

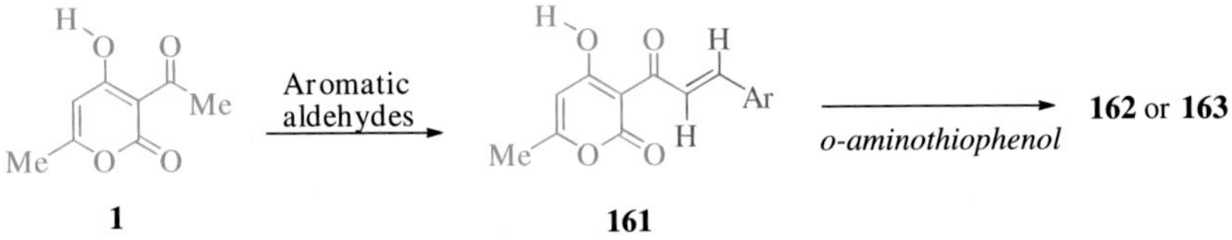

SCHEME 3.22   Dihydro-1, 5-benzothiazepines and dihydrobenzothiazines.

In the first step, **DHA** reacts with aromatic aldehyde to afford α,β-unsaturated ketones (**161**). In the second step, α,β-unsaturated ketones (**161**) reacts with *o*-ATP to yield two different compounds, namely dihydro-1,5-benzothiazepines (**162**) and dihydrobenzothiazines (**163**) (Scheme 3.22).

3.7   SYNTHESIS OF CINNAMOYL DERIVATIVES OF DHA

Two methods are available to synthesize cinnamoyl derivatives (**165**). The conventional method is one and the microwave method is the other [111] (Scheme 3.23).

3.7.1   Conventional Method

The **DHA** was reacted with different aldehydes (**164**) in chloroform in the presence of pyridine and piperidine as catalysts. Then, the mixture was heated by

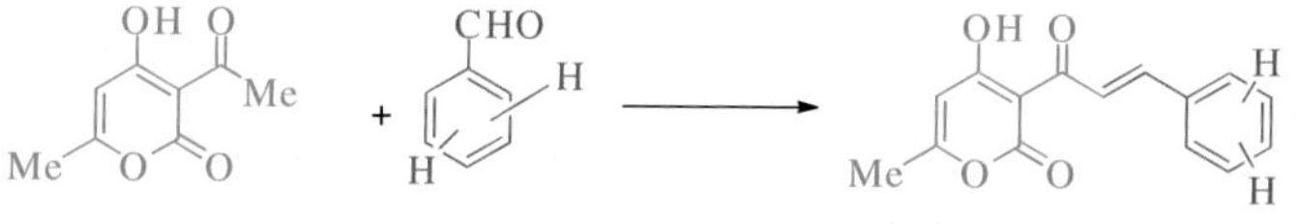

SCHEME 3.23   Synthesis of cinnamoyl derivatives of DHA.

using a conventional electrical heated jacket for about 6–9 hours, evaporated, and the pure solid product (**165**) was extracted.

3.7.2   Microwave Method

The **DHA** was reacted with different aldehydes (**164**) in neutral alumina in the presence of pyridine and piperidine as catalysts. Then, the mixture was heated by using microwaves for about 2–10 minutes, the solvent was evaporated, and the final product was extracted.

By comparing both the procedures, microwave irradiation of **DHA** and benzaldehyde derivatives (**Procedure B**) gave higher yields than the Conventional **Procedure A**.

3.8   SYNTHESIS OF VARIOUS DHA DERIVATIVES BY USING THE MULTICOMPONENT APPROACH

3.8.1   Introduction

Multicomponent reactions are very important reactions, in which by using a one-pot reaction more than three different reactants directly get converted into products. To synthesize new heterocyclic compounds in a single step, this is one of the best ways (Fig. 3.1). It is a very powerful tool in drug discovery and combinational chemistry [112].

Multicomponent reactions have been known for over 150 years. The first reported multicomponent reaction was Strecker's synthesis [113] of α-amino cyanides (**166**) in 1850 (Scheme 3.24).

Another example of multicomponent reactions is Hantzsch's pyrrole synthesis [114]. In this, pyrroles (**167**) are synthesized (Scheme 3.25).

Different dehydroacetic acid derivatives have been synthesized by many researchers using the multicomponent approach. Here, we will discuss some of those reactions.

3.8.2   Imidazol Derivatives of DHA

The reaction of the bromo derivative of **DHA** (**168**) with different aromatic aldehydes (**169**), phenylmethanamine (**170**), and ammonium acetate (**171**) in

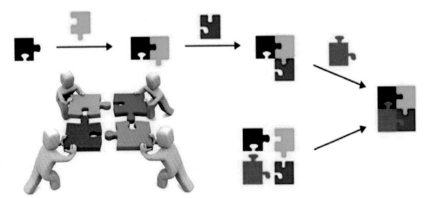

FIGURE 3.1 Multicomponent reactions. *Alvim HGO, da Silva Jr EN, Neto BAD. What do we know about multicomponent reactions? Mechanisms and trends for the Biginelli, Hantzsch, Mannich, Passerini and Ugi MCRs, RSC Adv 2013. doi:10.1039/C4RA10651B.*

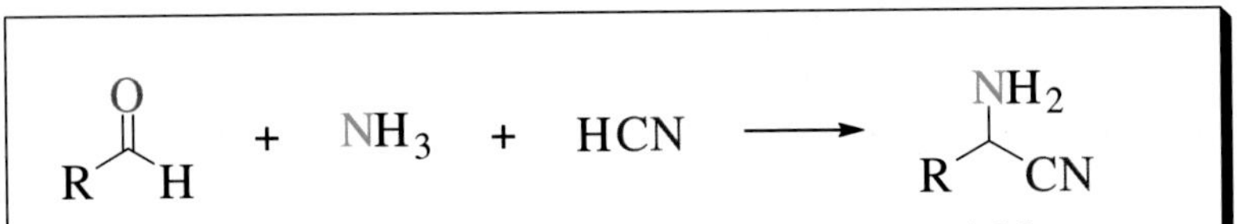

$$\underset{R}{\overset{O}{\|}}{\underset{H}{}} + NH_3 + HCN \longrightarrow \underset{R}{\overset{NH_2}{\|}}\underset{CN}{}$$

166

SCHEME 3.24 Strecker synthesis of α-amino cyanides.

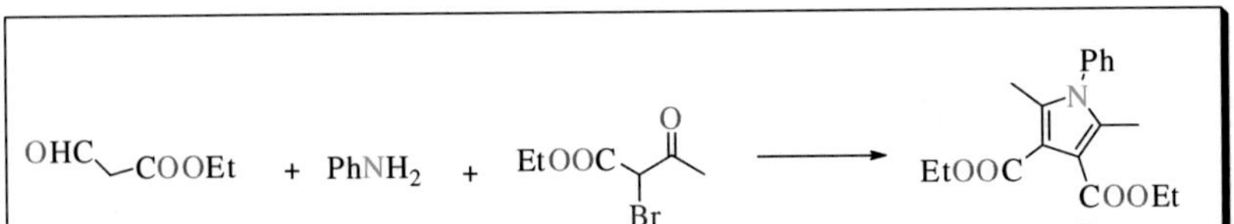

$$OHC\text{-}COOEt + PhNH_2 + EtOOC\underset{Br}{\overset{O}{\|}} \longrightarrow \text{EtOOC}\underset{COOEt}{\overset{Ph}{\|}}$$

167

SCHEME 3.25 Synthesis of pyrroles by using MCRs.

the presence of absolute alcohol resulted in the formation of imidazole derivatives of **DHA** (**172a–l**) in good yields [115] (Scheme 3.26).

The mechanism of the above reaction is explained here. In the first step, the condensation reaction of the amine compound (**170**) and aldehyde (**169**) resulted in an imine compound. The resulting imine compound was added to the bromodehydroacetic acid by following the nucleophilic addition reaction and formed the iminium intermediate, which was followed by the addition of ammonia and resulted in another intermediate. In the next step, the dihydroimidazole compound was formed by the intramolecular condensation of amino functionality of the intermediate with the adjacent carbonyl group, which was oxidized to yield the imidazole derivatives (**172a–l**) (Scheme 3.27).

Imidazole derivatives (**172a–l**) are listed in Table 3.4.

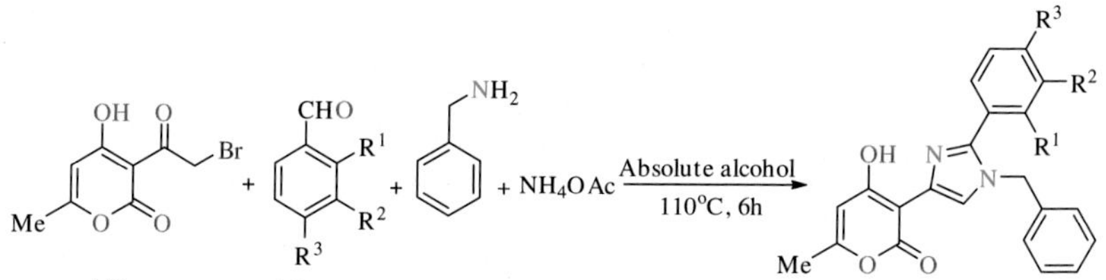

SCHEME 3.26 Synthesis of imidazoles my using MCRs.

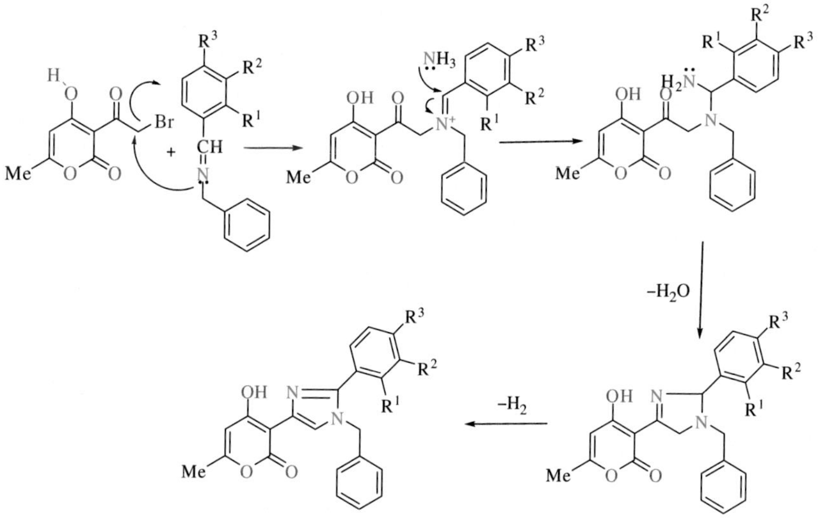

SCHEME 3.27 Mechanism of synthesis of imidazoles my using MCRs.

3.8.3 Thiadiazine Derivatives of DHA

Here the synthesis of thiadiazine derivatives using a solvent-free method is explained. The reaction of bromo DHA (**168**) with thiocarbohydrazide (**173**) and various carbonyl compounds (**174**) on stirring at 37°C for up to 30 minutes resulted in the formation of thiadiazine derivatives of **DHA** [116] (**175a–k**) (Scheme 3.28). The mechanism of the reaction was described as thiocarbohydrazide S-acylation with compound (**168**) and cyclization (intramolecular) to give DHA substituted hydrazinothiadiazines. This intermediate again underwent further condensation with carbonyl compounds and gave Schiff bases (**175a–k**) (Scheme 3.29).

The derivatives of the thiadiazines are listed in Table 3.5.

3.8.4 Thiazolyl-Pyrazolone Derivatives of DHA

The reaction of the bromo derivative of **DHA** (**168**), thiosemicarbazide (**173**), and ethyl 2-(2-phenylhydrazono)-3-oxobutanoate (**176**) under reflux in

TABLE 3.4 Imidazole Derivatives (172a–l)

Entry	R_1	R_2	R_3
172a	H	H	OCH_3
172b	H	OCH_3	OCH_3
172c	H	H	H
172d	H	H	OH
172e	OH	OCH_3	H
172f	OH	H	H
172g	H	H	Cl
172h	H	H	$N(CH_3)_2$
172i	H	H	Br
172j	H	NO_2	H
172k	H	OCH_3	OH
172l	OH	H	OH

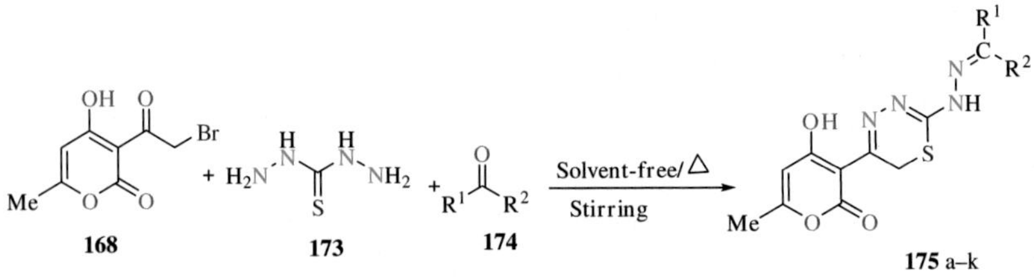

SCHEME 3.28   Synthesis of thiadiazines by using MCRs.

anhydrous ethanol gave the final products of thiazolyl-pyrazolone derivatives of **DHA** [117] (**177a–k**) (Scheme 3.30).

The derivatives of the thiazolyl-pyrazolone derivatives of **DHA** are listed in Table 3.6.

The mechanism of the formation of the thiazolyl-pyrazolone derivatives of **DHA** is shown in Scheme 3.31.

3.8.5   Pyridine Derivatives of Dehydroacetic Acid via Hantzsch Reaction

The reaction of **DHA** with 4-nitrobenzaldehyde (**178**) and ammonium acetate (**171**) using catalytic amounts of **CAN** (ceric ammonium nitrate) at ambient

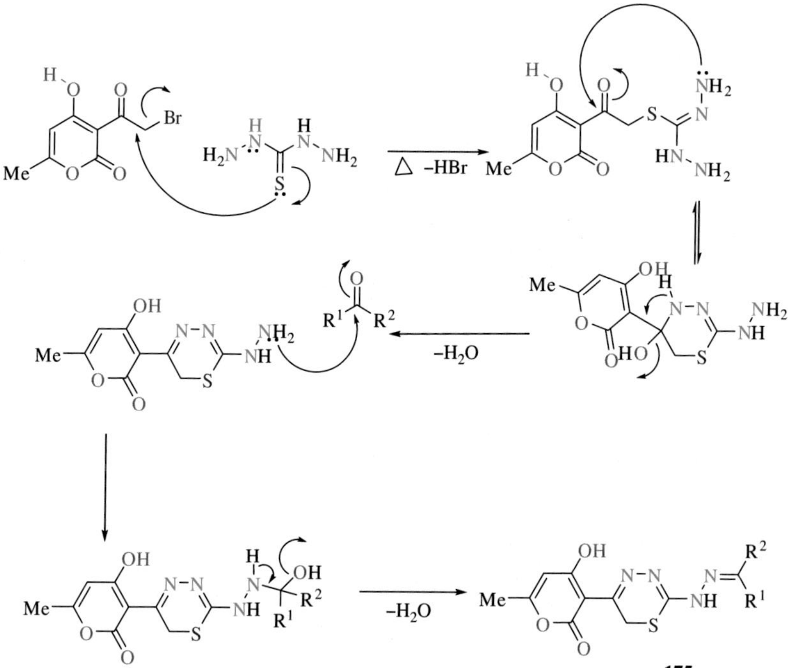

SCHEME 3.29 Mechanism of synthesis of thiadiazines by using MCRs.

TABLE 3.5 Derivatives of the Thiadiazines

Entry	R_1	R_2
175a	H	p-Tolyl
175b	H	p-Anisyl
175c	H	o-Hydroxy phenyl
175d	H	Phenyl
175e	Me	Me
175f	Me	Ph
175g	Me	p-Tolyl
175h	Me	p-Anisyl
175i	Me	p-Chloro phenyl
175j	Me	Et
175k	$R_1 = R_2 = $ Cyclohexylidene	

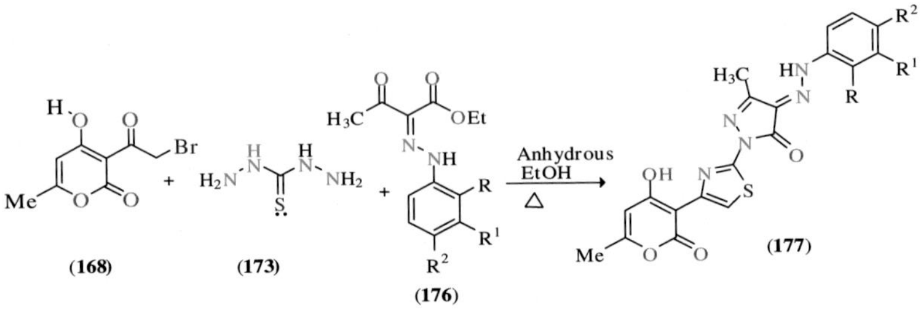

SCHEME 3.30 Synthesis of thiazolylpyrazolones by using MCRs.

TABLE 3.6 Derivatives of the Thiazolyl-Pyrazolone

Entry	R	R$_1$	R$_2$
a	H	H	H
b	H	H	CH$_3$
c	CH$_3$	H	H
D	H	H	NO$_2$
E	H	NO$_2$	H
F	NO$_2$	H	H
G	OCH$_3$	H	H
H	H	OCH$_3$	H
I	H	H	OCH$_3$
J	H	H	Cl
K	COOH	H	H

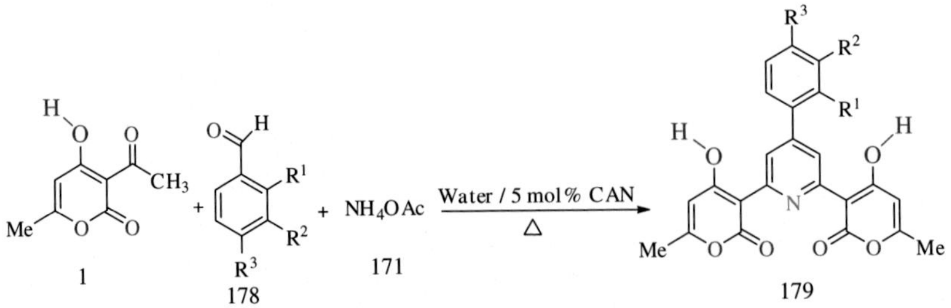

SCHEME 3.31 Mechanism of synthesis of thiazolylpyrazolones.

temperature in an aqueous medium gave 2,4,6-tri-substituted pyridine derivatives (**179**) in good yields [118] (Scheme 3.32). The reaction was carried out
without any catalyst. In that case, yields of the 2,4,6-tri-substituted pyridine
derivatives were very poor (30%–40%). However, using CAN as a catalyst, the
reaction gave a yellow product with 90% yield in 3 hours.

The pyridine derivatives of dehydroacetic acid are listed in Table 3.7.

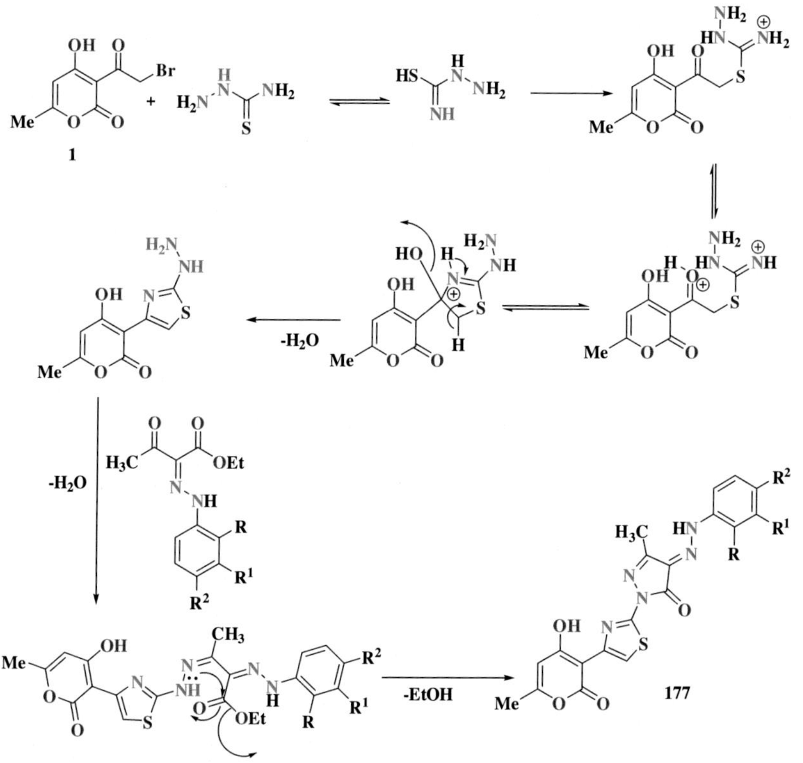

SCHEME 3.32 Synthesis of pyridine derivatives of DHA via Hantzsch reaction.

TABLE 3.7 Pyridine Derivatives of Dehydroacetic Acid

Compound	R₁	R₂	R₃
4a	H	H	NO$_2$
b	H	OCH$_3$	OCH$_3$
c	H	H	Br

(*Continued*)

TABLE 3.7 Pyridine Derivatives of Dehydroacetic Acid (Continued)

Compound	R_1	R_2	R_3
d	H	H	Cl
e	H	H	CH_3
f	H	H	OCH_3
g	H	H	$N(CH_3)_2$
h	H	H	OH
i	OH	H	H
j	H	NO_2	H
k	2-Hydroxy-1-naphthaldehyde		

Chapter 4

Dehydroacetic Acid–Metal Complexes

Gudala Satish

National Institute of Technology Raipur, Raipur, Chhattisgarh, India

4.1 INTRODUCTION AND IMPORTANCE OF FE COMPLEXES

Many of us have heard the regular proclamation from our seniors that they have gold in their teeth, silver in their hair, and lead in their bones. Weird as it may appear at first sight, every one of the three metals can have an impact on living system. These metals usually bind to ligands (atoms, ions, or molecules which donate electrons to the metal) and form important biologically active metal complexes [119]. Normally these ligands are organic compounds and are called chelants, chelators, and chelating agents (Fig. 4.1).

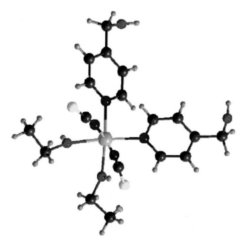

FIGURE 4.1 Crystal structure of a metal complex. *Karkas MD, Akermark B. Water oxidation using earth-abundant transition metal catalysts: opportunities and challenges, Dalton Trans 2016:45;14421–61.*

Dehydroacetic Acid and Its Derivatives. DOI: http://dx.doi.org/10.1016/B978-0-08-101926-9.00004-3
© 2017 Elsevier Ltd. All rights reserved.

All these complexes are neutral or charged. Different metals play key parts in living systems and they can simply lose electrons to finally form positive ions. They tend to be soluble in natural fluids. These metal particles are electron poor, though the majority of the biomolecules, e.g., deoxyribonucleic acid (DNA) and proteins are electron rich. The attraction between these opposite charges leads to interaction and binding between metal ions and biological molecules.

Iron can perform a wide variety of tasks, e.g., the transport of oxygen to every one of the body's parts. Hemoglobin (Fig. 4.2) is a protein containing heme as the prosthetic group, it contains iron in its structure. It binds with oxygen and transports it to different body tissues [120].

Iron bleomycin (**180**) was the initially reported natural product to cleave DNA in an oxidative pathway. Its significance was perceived because of its novel and broad spectrum antitumor properties [121,122].

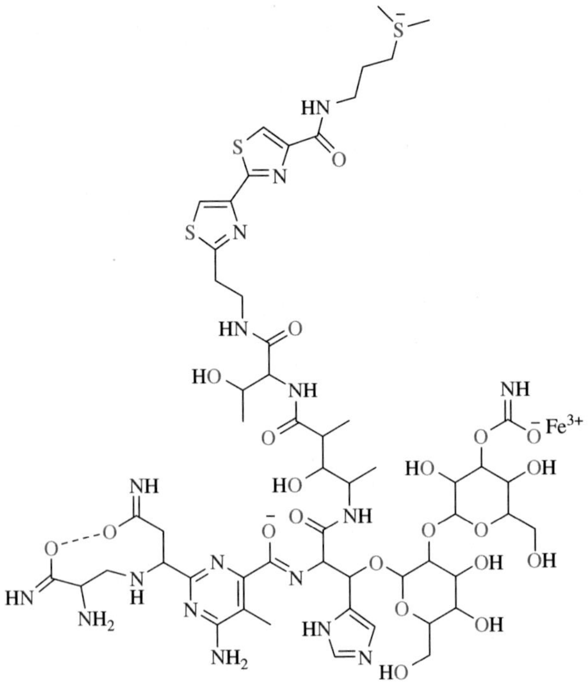

180

4.2 IMPORTANCE OF PT COMPLEXES

The interactions of metal complexes with DNA have been widely examined for their use as probes in the formation of DNA structure. They have potential

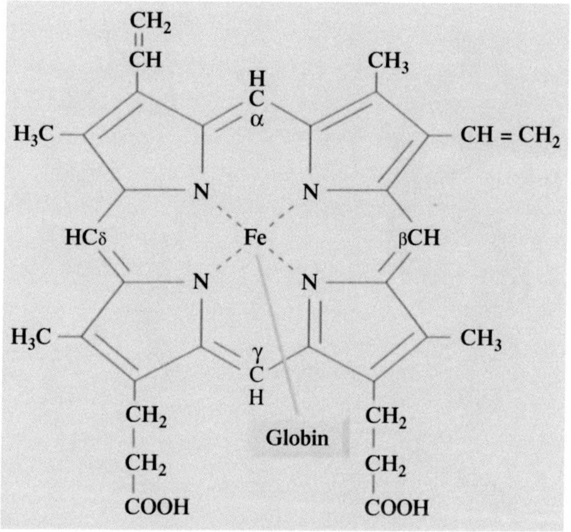

FIGURE 4.2   Structure of hemoglobin. *Adams UI, Abdullahi U, Saliha BS, Happiness UI. Color matching estimation of iron concentrations in branded iron supplements marketed in Nigeria. Adv Anal Chem 2012:2;16–23.*

applications in chemotherapy by inhibiting the DNA gyrase [123–125]. The immense victories achieved with platinum-based antitumor agents, including cis-platin (**181**), carboplatin (**182**), and oxaliplatin (**183**), have empowered the improvement of metal-based drugs.

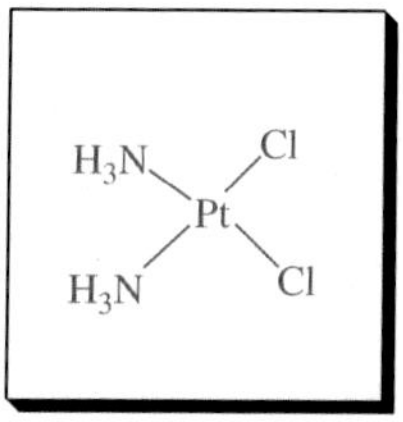

181 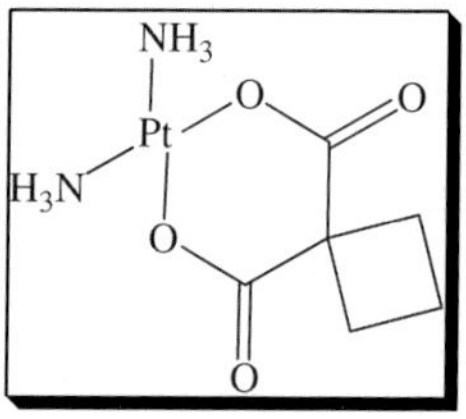   **182** 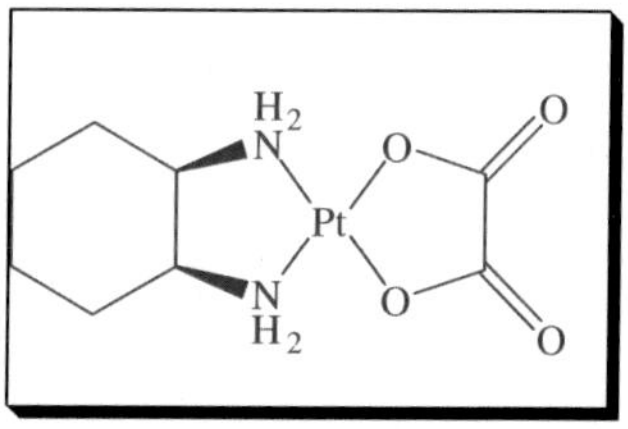   **183**

We will get a clear idea about the interaction of the transition metal complex cis-platin with DNA after observing its mechanism. Cis-platin is responsible for the platination of DNA present inside the cell by cross-connecting at the N7 position of the guanine base (the most oxidizable and electron rich site) (Fig. 4.3). The formation of cisplatin DNA adducts causes alterations which result in the inhibition of DNA replication [126,127].

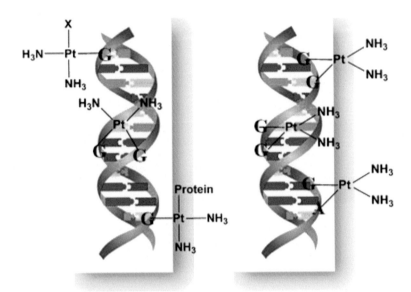

FIGURE 4.3 DNA binding of cis-platin. *Williams DR. The metals of life. London: Van Nostrand Reinhold, London; 1971 OR Zivadin D, Bugarcic, JB, Biljana P, Stephanie H, Rudi van Eldik. Mechanistic studies on the reactions of platinum (II) complexes with nitrogen and sulfur-donor biomolecules. Dalton Trans 2012;41:12329.*

4.3 IMPORTANCE OF CU COMPLEXES

Copper is essential for the proper functioning of copper-dependent enzymes, including **cytochrome C** oxidase (production of energy), dopamine hydroxylase (catecholamine production), superoxide dismutase (antioxidant protection), tyrosinase (pigmentation), clotting factor V (blood clotting), lysyl oxidase (collagen and elastin formation), and ceruloplasmin (copper transport, iron metabolism, and antioxidant protection) [128]. Copper (II) has proved to be useful in infections, e.g., tuberculosis, gastric ulcers, and rheumatoid joint pains [129,130], and it has been found to play a significant role in biological processes, e.g., **Cu (DMG)$_2$** (**184**) shows high activity against cancer and enhances the life span to the extent of 20%–30%.

184

4.4 IMPORTANCE OF RU AND ZN COMPLEXES

In the most recent decades, extensive research has been undertaken on ruthenium complexes, such as NAMI-A (**185**) and KP1019 (**186**). These complexes have been also entered into clinical trials [131,132].

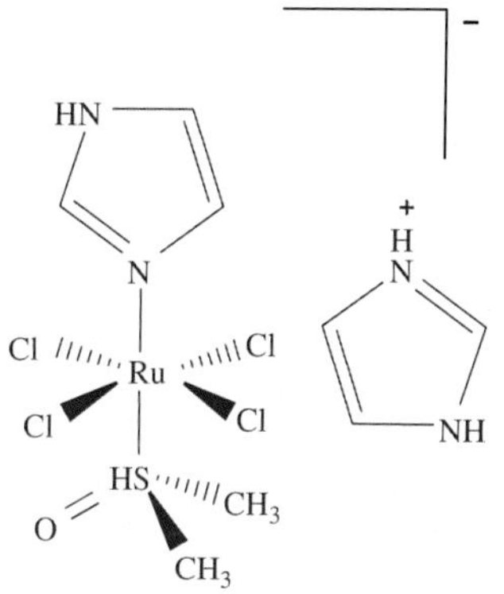

185

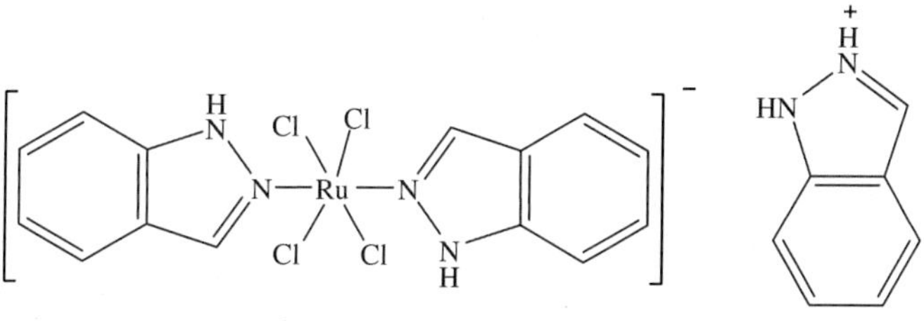

186

Zinc is a constituent of carbonic anhydrase enzyme (**187**) which is involved in the conversion of CO_2 to carbonic acid in plants. The 2-[(dimethylamino) methyl] phenyl gold (III) complex (**188**) has also been proven to be an antitumor agent against human cancers [133].

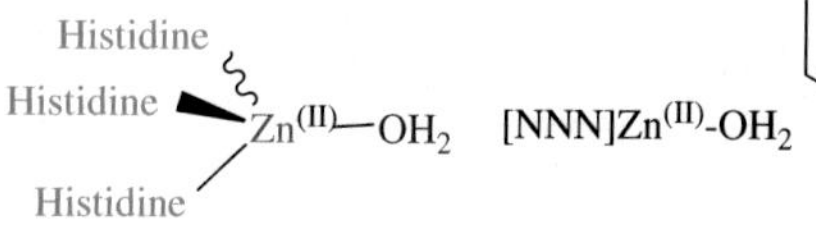

187 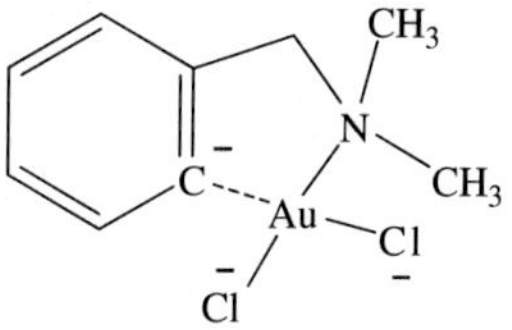   **188**

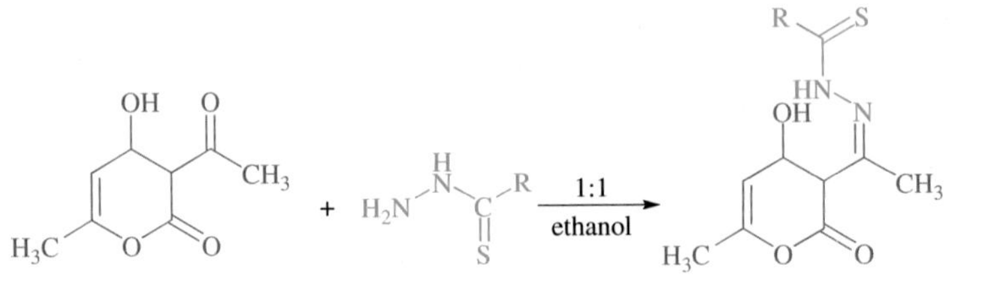

SCHEME 4.1 Synthesis of DHA schiff bases from methyl hydrazinecarbodithioate and thiosemicarbazide.

4.5 PD (II) AND NI (II) CHELATES OF DHA

The reaction of DHA (**1**) with methyl hydrazinecarbodithioate (**189**)/thiosemicarbazide (**190**) in boiling ethanol afforded the corresponding Schiff bases (Z)-methyl 2-(1-(4-hydroxy-6-methyl-2-oxo-3,4-dihydro-2H-pyran-3-yl) ethylidene)hydrazinecarbodithioate (**DAE (191)**) and (Z)-2-(1-(4-hydroxy-6-methyl-2-oxo-3,4-dihydro-2H-pyran-3-yl)ethylidene)hydrazinecarbothioamide (**DATS (192)**), respectively (Scheme 4.1).

These **DAE** and **DATS** compounds were reacted with nickel (II) acetate and given crystalline chelates (brown) of [Ni (DATS-2H)]$_2$ and [Ni (DAE-2H)]$_2$ (**193**), respectively (Scheme 4.2).

DATS-2H and DAE-2H are the dinegative anionic forms of the ligands DATS and DAE, respectively.

4.6 SYNTHESIS OF MANGANESE COMPLEXES FROM DHA

Mn(dha)$_2$(CH$_3$OH)$_2$ complexes were synthesized by mixing **DHA** and Mn (OAc)$_2$ 4H$_2$O in methanol. Mn(dha)$_2$(CH$_3$OH)$_2$ can also be synthesized from the reaction of two equivalents of **DHA** with Mn (III) acetate in the presence of air [134]. The crystallographic drawing of [Mn(dha)$_2$ [Mn(dha)$_2$(CH$_3$OH)$_2$] was taken by using the Oak Ridge Thermal-Ellipsoid Plot Program (**ORTEP**). Fig. 4.4 shows the ORTEP drawing of Mn(dha)$_2$(CH$_3$OH)$_2$.

4.6.1 Mn(dha)$_2$(CH$_3$OH)$_2$ X-Ray Crystal Structure

In each unit cell, there are two Mn(dha)$_2$(CH$_3$OH)$_2$ molecules. The Mn (II) is bonded to two oxygen atoms from the coordinated methanol molecules and four oxygen atoms from two **DHA** ligands. The Mn(dha)$_2$(CH$_3$OH)$_2$ coordination geometry is octahedral with two methanol molecules filling the two axial sites and with four oxygen atoms of **DHA** ligands occupying the four equatorial positions. From the calculated least square plane ring structure, a planar six-membered ring can be formed by the **DHA** ligand and Mn (II). The Mn (II) will be in center of the square plane formed by four oxygen atoms from the two DHA molecules. The position of the two **DHA** ligands is in the trans configuration without a significant distortion from the equatorial plane.

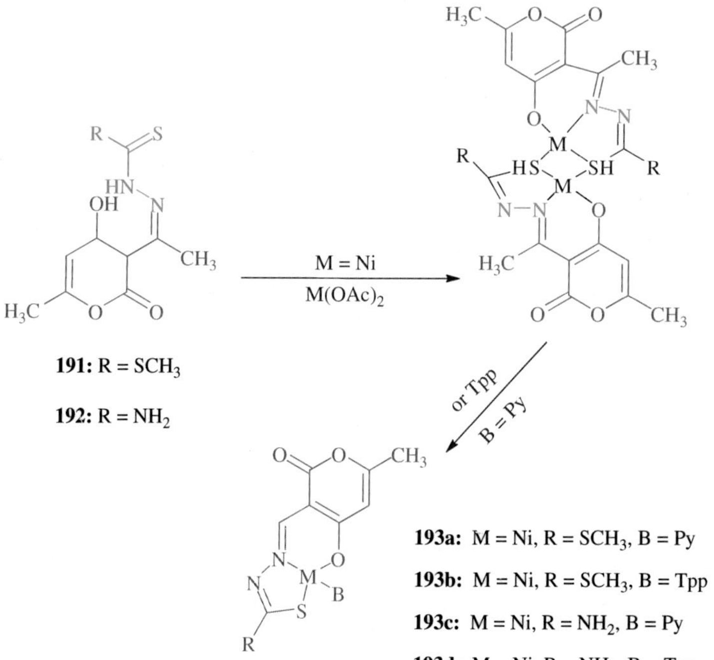

191: R = SCH₃

192: R = NH₂

M = Ni

M(OAc)₂

or Tpp

B = Py

193a: M = Ni, R = SCH₃, B = Py

193b: M = Ni, R = SCH₃, B = Tpp

193c: M = Ni, R = NH₂, B = Py

193d: M = Ni, R = NH₂, B = Tpp

SCHEME 4.2   Synthesis of Pd (II) and Ni (II) chelates of DHA.

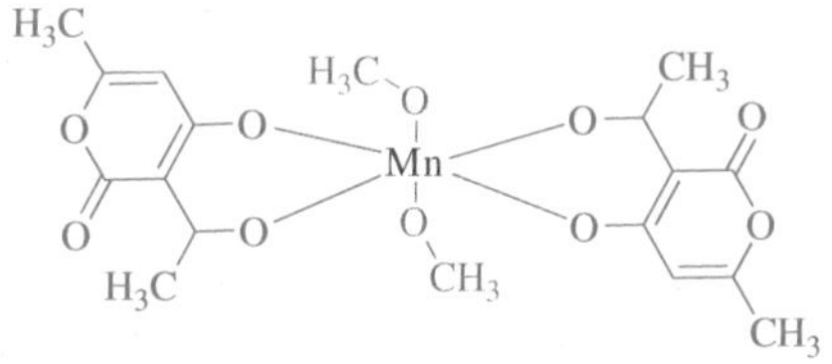

FIGURE 4.4   Structure of Mn–DHA complex. *Hsieh W, Zaleski CM, Pecoraro VL, Fanwick PE, Liu S. Mn(II) complexes of monoanionic bidentate chelators: X-ray crystal structures of Mn(dha)₂(CH₃OH)₂ (Hdha = dehydroacetic acid) and [Mn(ema)₂(H₂O)]₂·2H₂O (Hema = 2-ethyl-3-hydroxy-4-pyrone). Inor Chim Acta 2006;359:228–36.*

4.7   SYNTHESIS OF RUTHENIUM (II) AND RUTHENIUM (III) COMPLEXES WITH DHA

4.7.1   Synthesis of [RuCl (CO) (PPh₃)₂(L)] Complex

DHA-metal complex [RuCl (CO) (PPh₃)₂(L)] (**195**) can be prepared in two steps. In the first step, the starting complex [RuHCl (CO) (PPh₃)₃] (**194**) was prepared by using previous literature [135–139]. Then it was suspended in a mixture of benzene and dry ethanol. In the second step, **DHA** ligand (**1**) is

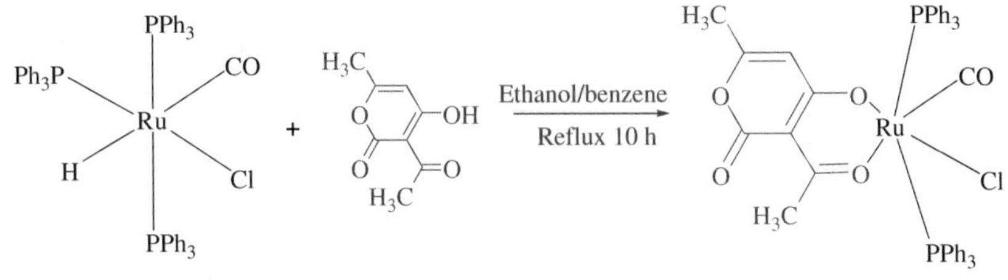

(194)   **(1)**   **(195)**

SCHEME 4.3   Synthesis of [RuCl (CO) (PPh3)2(L)] metal complex.

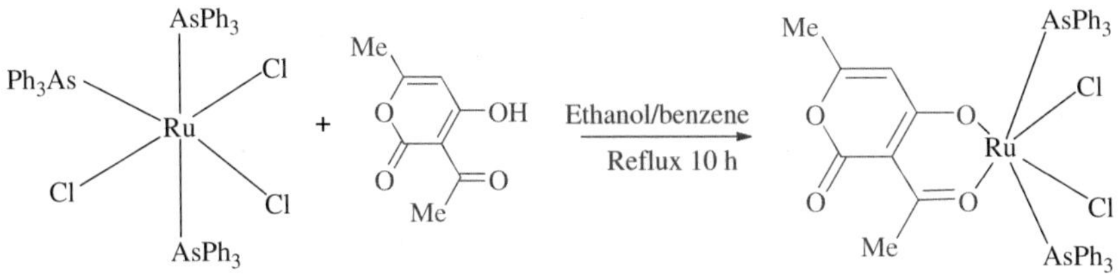

196   **1**   **197**

SCHEME 4.4   Synthesis of [RuCl2 (AsPh3)2 (L)] metal complex.

added to the above complex and refluxed and then it is cooled and filtered. The complex can be separated out by the solvent evaporation up to a small volume. The compound can be recrystallized from a CHCl₃/petroleum ether (1:1) mixture (Scheme 4.3).

4.7.2   Synthesis of [RuCl₂ (AsPh₃)₂L]

As mentioned earlier, the complex [RuCl (CO) (PPh₃)₂ (L)] was synthesized as described in the literature [121–125]. Here the synthesis of complex [RuCl₂ (AsPh₃)₂ (L)] (**197**) was synthesized by the suspension of complex [RuCl₃ (AsPh₃)₃] (**196**) in the mixture of benzene and dry ethanol. To this, **DHA** ligand was added and the mixture was refluxed. Then the solution was filtered and allowed to stand for 3 days leading to the appearance of microcrystalline product [140]. Addition of dimethyl formamide (DMF) to the solution yielded single crystals which are appropriate for X-ray diffraction studies (Scheme 4.4).

The ligand **DHA** and its complexes possess antibacterial activity against gram-positive bacteria, such as *Staphylococcus epidermidis* [141] (Fig. 4.5A) and *Staphylococcus aureus* [142] (Fig. 4.5B), and gram-negative bacteria, such as *Shigella sonnei* [143] (Fig. 4.6A) and *Escherichia coli* [144] (Fig. 4.6B).

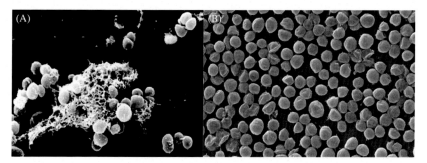

FIGURE 4.5   (A) Scanning electron microscopy. (B) *S. aureus* of a naturally attached *S. epidermidis*. *(A) Natascha K, Rosana BRF, Ana PFN, Ulisses GCL, Fernando CSF, Ana LM-G, et al. Cell surface hydrophobicity and slime production of Staphylococcus epidermidis Brazilian isolates. Curr Microbiol 2003;46:280–6. (B) Di Ciccio P, Vergara A, Festino AR, Paludi D, Zanardi E, Ghidini S, Ianieri A, Biofilm formation by Staphylococcus aureus on food contact surfaces: relationship with temperature and cell surface hydrophobicity. Food Contr 2014. doi: 10.1016/j.foodcont.2014.10.048.*

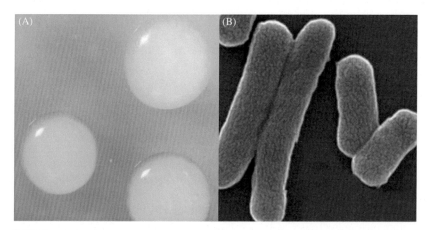

FIGURE 4.6   (A) Appearance of *S. sonnei*. (B) Scanning electron microscopy colonies after 48 h on M9GRN (M9 images of *E. coli* cells containing riboflavin and nicotinic acid). *(A) Deldar AA, Yakhchali B. The influence of riboflavin and nicotinic acid on Shigella sonnei colony conversion, Iran. J Microbiol 2011:3;13–20. (B) Agnieszka Z-B, Pawel M, Anna K, Krzysztof S, Ewaryst M, Marta JF, et al. Synergistic action of Galleria mellonella anionic peptide 2 and lysozyme against Gram-negative bacteria. Biochimica et Biophysica Acta 2012; 1818:2623–35.*

4.8   SYNTHESIS OF CU (II), CO (II), NI (II), AND CD (II) COMPLEXES USING SODIUM SALT OF DHA

A very important series of newly mixed ligand complexes of cadmium (II), cobalt (II), copper (II), and nickel (II) mixed ligand complexes have been synthesized by using methoxybenzylidine aminotriazolethione (**MBT**) (**198**), chlorobenzylidineaminotriazolethione (**CBT**) (**199**), nitrobenzylidineaminotriazolethione (**NBT**) (**200**), and sodium salt of dehydroacetic acid (**NaDHA**) [145].

4.8.1 Synthesis of Mercaptotriazole Ligands

MBT, **CBT**, and **NBT** are synthesized by using previous literature [146–148]. The structures of ligands are mentioned as follows.

(198–200)

(1)

Where Ar is

3-benzyl-4-[(2-methoxybenzylidene) amino]-1H-1, 2, 4-triazolo-5-thione (MBT)(198)

3-benzyl-4-[(4-chlorobenzylidene) amino]-1H-1, 2, 4-triazolo-5-thione (CBT)(199)

3-benzyl-4-[(4-nitrobenzylidene) amino]-1H-1, 2, 4-triazolo-5-thione (NBT)(200)

Structures of Ligands

4.8.2 Synthesis of Cu (II) Mixed Ligand Complexes

To the **CBT**, **MBT**, or **NBT** ligand solution, a solution of copper chloride is to be added drop wise with continuous stirring in one direction. After formation of the precipitate, the **NaDHA** ligand should be added and refluxed to obtain the final products. Wash the compound with dry methanol and dry it using a vacuum over anhydrous $CaCl_2$.

4.8.3 Synthesis of Ni (II), Cd (II), and Co (II) Mixed Ligand Complexes

To the mixture of $CoCl_2 \cdot 6H_2O/NiCl_2 \cdot 6H_2O$ or $CdCl_2 \cdot 2.5H_2O$ and sodium acetate (1:2) in methanol, a solution of **CBT**, **MBT**, or **NBT** (25 mL hot methanol and 1 mmol ligand) is added drop wise with continuous stirring in one direction. The further process is the same as described earlier to get complexes (**201a–d**).

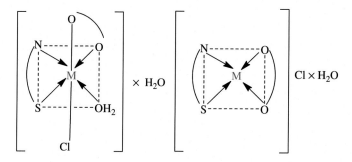

201 a-d: M = Cd (II), Co (II), Cu (II) or Ni (II); x = 0 or 1

4.9 SYNTHESIS OF FE (III), NI (II), CU (II), CO (II), MN (II), AND COMPLEXES BY USING *O*-PHENYLENEDIAMINE, DHA, AND *P*-CHLOROBENZALDEHYDE

Tridentate Schiff base ligands were synthesized from DHA, *o*-phenylenediamine, and *p*-chlorobenzaldehyde. These Schiff base ligands are used to synthesize solid complexes of Fe (III), Ni (II), Cu (II), Co (II), and Mn (II) [149].

4.9.1 Synthesis of Ligand

The ligand (**205**) was prepared by taking a solution of **DHA** (**1**), *o*-phenylenediamine (**202**) in ethanol and refluxed. The resulting compound (*E*)-3-(1-(2-aminophenylimino)ethyl)-4-hydroxy-6-methyl-2*H*-pyran-2-one (**203**) was reacted with *p*-chlorobenzaldehyde (**204**) in dry ethanol and refluxed. The final compound (4-hydroxy-3(1-(2-(benzylideneamino)-phenylimino)-ethyl)-6-methyl-2*H*-pyran-2-one) (**196**) was filtered and dried in vacuum over CaCl₂ and recrystallized in ethanol (Scheme 4.5).

4.9.2 Synthesis of Metal Complexes

To the hot solution of the ligand, a methanolic solution of metal chloride was added with constant stirring. The pH of the reaction mixture was adjusted to 7.5–8.5 by adding 10% alcoholic ammonia solution and then refluxed until a solid complex appeared. The resulting metal complex was filtered off in hot conditions. The final compounds were washed with petroleum ether or hot methanol and then dried using a vacuum desiccator over calcium chloride.

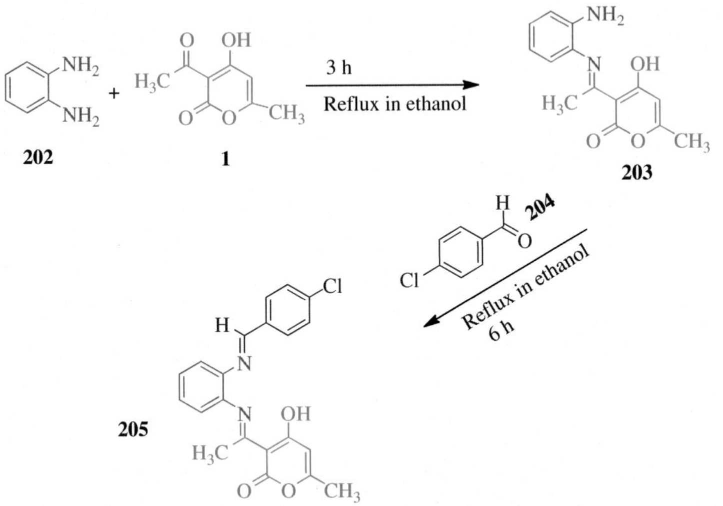

SCHEME 4.5 Synthesis of ligand from DHA, o-phenylenediamine and aldehyde.

4.10 SYNTHESIS OF FE (III), CO (II), NI (II), MN (II), AND CU (II) COMPLEXES BY USING DHA, 4-METHYL-*O*-PHENYLENE DIAMINE, AND SALICYLIC ALDEHYDE

Fe (III), Co (II), Ni (II), Mn (II), and Cu (II) complexes were prepared by using asymmetric tetra dentate Schiff base ligand (**206**) which is derived from DHA 4-methyl-*o*-phenylene diamine and salicylic aldehyde [143].

The tetra dentate asymmetric Schiff base ligand can be prepared via stepwise approach. In the first step the mono-Schiff base compound is prepared by refluxing DHA and 4-methyl-*o*-phenylenediamine in dry ethanol for 3 hours. The resulting mono-Schiff base should be refluxed with salicylic aldehyde to prepare the asymmetric ligand, namely, 4-hydroxy-3-(1-{{2-[(2-hydroxyben-zylidene) amino]-4-methylphenyl} imino} ethyl)-6-methyl-2*H*-pyran-2-one (H$_2$L) (**206**). The formed asymmetric Schiff base has to be cooled to room temperature and collected by filtration followed by recrystallization in ethanol.

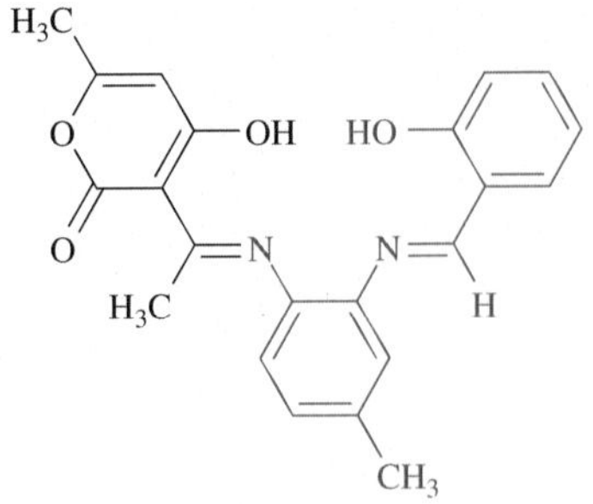

206

4.10.1   Synthesis of Metal Complexes

To a hot methanolic solution of the ligand, a methanolic solution of a metal chloride under constant stirring should be added. The pH of the reaction mixture is adjusted to 7.5–8.5 by adding 10% alcoholic ammonia solution and the solution is refluxed until solid complexes appear. The formed solid metal complexes (**207a–b, 208a–c**) are filtered off under hot conditions. All the complexes are washed with hot methanol, petroleum ether, and then dried over anhydrous CaCl$_2$ in a vacuum desiccator.

207a,b: M (II) = Cu(II) or Ni(II)   **208a–c:** M = Fe(III), Mn(II) or Co(II).

4.11   SYNTHESIS OF MN (II), CO (II), NI (II), AND CU (II) COMPLEXES FROM PHENYL ANILINE–DHA LIGAND

4.11.1   Preparation of the Ligands

The ligands DHA–phenyl aniline (**L^1**), DHA-4-chlorophenylaniline (**L^2**), DHA-4-iodophenyl aniline (**L^3**), and DHA-4-ethoxy phenyl aniline (**L^4**) are prepared by mixing equimolar solutions of **DHA** and the corresponding amine in ethanol and refluxing the mixture for 4–5 hours. After the contents are cooled to room temperature and solid is separated out by washing and recrystallization.

4.11.2   Preparation of Metal Complexes

To a hot solution of ligand in methanol, the metal chloride in methanol is to be added drop wise. The pH of the solution is adjusted to the required value by adding alcoholic ammonia. The respective metal complexes (colored solids) will be separated by filtering in hot conditions and washing with hot methanol followed by petroleum ether and dried in vacuum [151]. The complexes of different metals will be precipitated at different pH.

4.12 SYNTHESIS OF CR (III), CO (II), OXO-VANADIUM (IV), MN (II), FE (II), CU (II), AND NI (II) COMPLEXES FROM 1,2-DIAMINOETHANE–DHA LIGAND

4.12.1 Synthesis of the Schiff Base (Ligand)

The reaction between DHA (**1**) and 1,2-diaminoethane (**209**) takes place in the presence of ethanol and refluxed in a water bath for 2 hours to give a yellow precipitate (ligand) (**210**). Then the ligand is filtered using reduced pressure and washed with alcohol. Recrystallize the compound from ethanol (Scheme 4.6).

4.12.2 Preparation of the Metal Complexes

The metal complexes (**211–214**) can be synthesized by refluxing the ligand and a hot ethanolic solution of metal chloride [except for Fe (II) and oxovanadium (IV)] in a water bath. By adjusting the pH to 7.5 the metal complexes can be separated. The compounds can be filtered with cyclohexanol and petroleum ether (60–80°C) and dried in air. In the case of oxovanadium (IV) and Fe (II) complexes, an aqueous ethanolic solution is to be used [152] (Schemes 4.7 and 4.8).

On the basis of spectral studies and magnetic data of Mn (II), oxo-vanadium (IV), Co (III), Cu (II), and Cr (III) complexes have octahedral geometries while the Ni (II) complex has a square planar geometry.

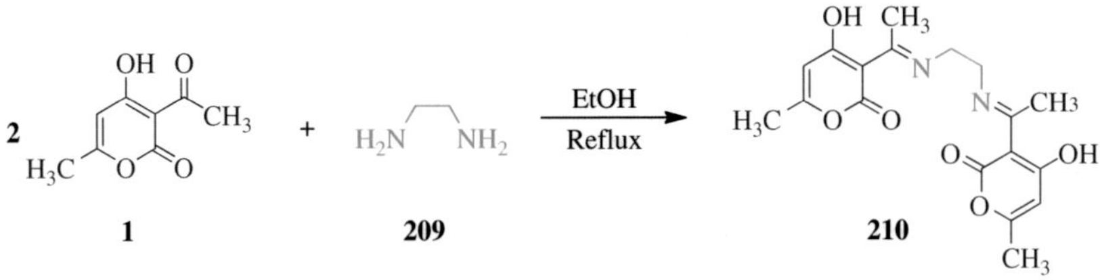

SCHEME 4.6 Synthesis of ligand from DHA and 1, 2-diaminoethane.

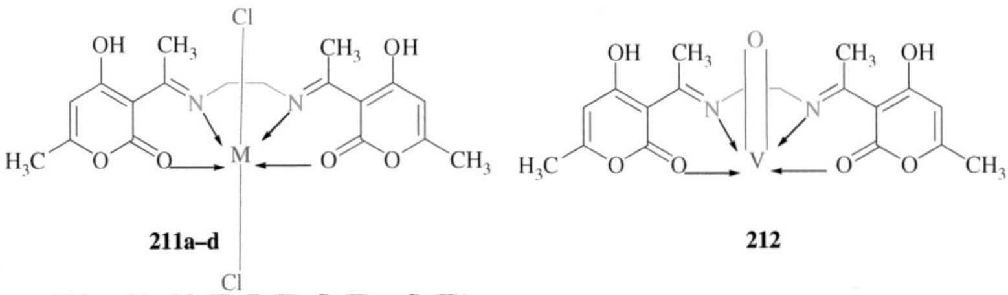

211a–d

(Where M = Mn(II), Fe(II), Co(II) or Cu(II))

212

SCHEME 4.7 Various metal complexes of DHA.

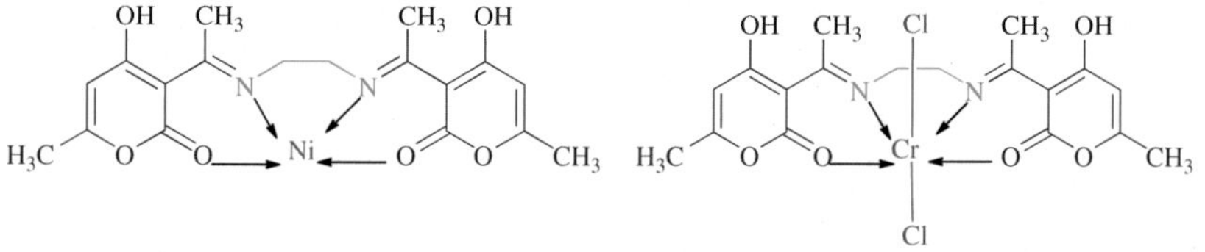

SCHEME 4.8 Ni and Cr metal complexes of DHA.

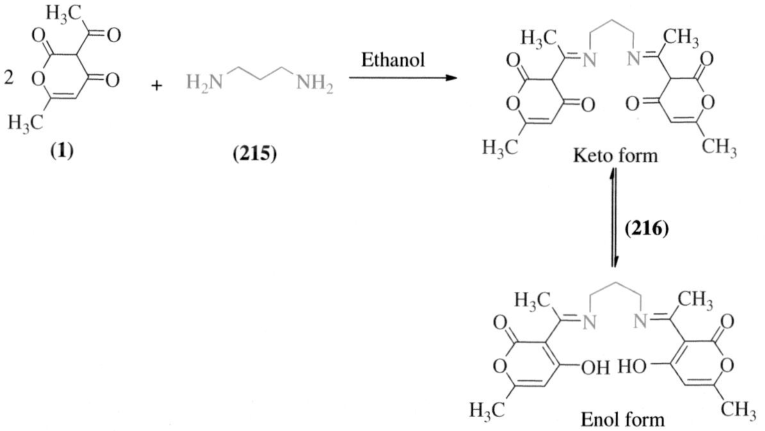

SCHEME 4.9 Synthesis of Schiff base ligand H2L.

4.13 SYNTHESIS OF TI (III), CR (III), FE (III), MN (III), ZR (IV), VO (IV), AND UO₂ (VI) COMPLEXES FROM DHA WITH 1,3-DIAMINOPROPANE

The complexes of Ti (III), Cr (III), Fe (III), Mn (III), Zr (IV), VO (IV), and UO$_2$ (VI) with Schiff base 3,3'-(1E,1'E)-1,1'-(propane-1,3-diylbis(azan-1-yl-1-ylidene))bis(ethan-1-yl-1-ylidene)bis(4-hydroxy-6-methyl-2H-pyran-2-one) (**H₂L**) (**216**) were derived from condensation of DHA (**1**) with 1,3-diaminopropane [153] (**215**).

4.13.1 Synthesis of H₂L Ligand

The Schiff base ligand **H₂L** (**216**) was synthesized by dissolving **DHA** (**1**) in ethanol taken in a round bottom flask, followed by the drop wise addition of 1,3-diaminopropane (**215**) with constant stirring and then cooled to room temperature (Scheme 4.9).

The Ti (III), Cr (III), Fe (III), Mn (III), VO (IV), and UO$_2$ (VI) complexes were synthesized by dissolving ligand in ethanol and heating at 80°C. To this hot solution, an ethanolic solution of the appropriate metal salt solution (in the case of VO (IV), DME was used in place of ethanol) was added drop by drop with continuous stirring and the resulting reaction mixture was further refluxed for 4–6 hours. The obtained solid product was washed thoroughly with ethanol and finally with petroleum ether. All these complexes were dried at room temperature over CaCl$_2$.

The reactions of Ti (III), Cr (III), Fe (III), Mn (III), Zr (IV), VO (IV), and UO$_2$ (VI) salts solution with **H$_2$L** resulted in the formation of complexes. All the complexes are colored solids, which decompose at high temperature.

4.14 SYNTHESIS OF NI (II), CO (II) COMPLEXES FROM METHANAMINE–DHA LIGAND

4.14.1 Synthesis of Schiff Bases

Schiff base (**218**) can be prepared by refluxing an equimolar mixture of amine (**217**) in ethanol and **DHA** [154]. The contents will be refluxed for 4 hours. All the formed solids are washed by ethanol, recrystallized, and dried in a vacuum desiccator. These ligands are stable to moisture and in air and they are soluble in alcohol, chloroform, dioxane, DMF, and dimethylsulfoxide (DMSO) (Scheme 4.10).

Metal complexes (**219a–c**) can be synthesized in a round bottomed flask by refluxing the ethanolic solution of metal chlorides and ligands in a 1:3 molar ratio and heated. This metal solution is added drop wise to a hot solution of ligands and refluxed. The complexes of different metals will be precipitated at different pH. Wash all the complexes with ammonia followed by petroleum ether at 40–60°C. Dry all the complexes in a desiccator over CaCl$_2$.

219

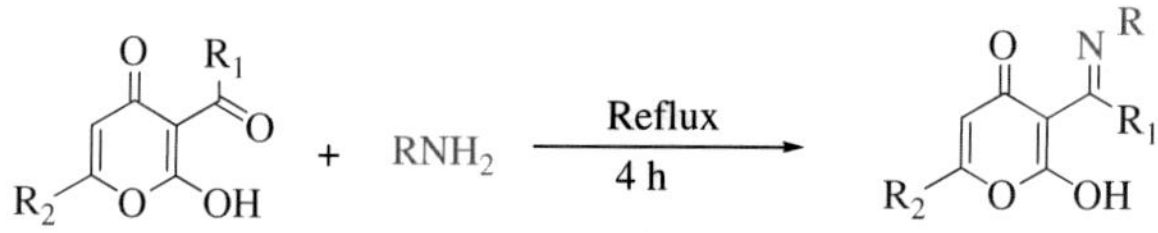

(1)   (217)   (218)

SCHEME 4.10   Synthesis of schiff base from DHA and amines.

TABLE 4.1 Various DHA Ligands

SR No.	R	R_1	R_2
218/219a	CH_3	CH_3	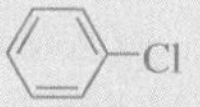
218/219b	CH_3	CH_3	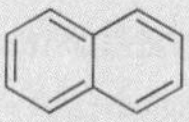
218/219c	CH_3	CH_3	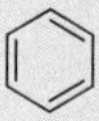

The derivatives of compounds **218** and **219** are given in Table 4.1.

All the metal complexes show antimicrobial activity and it is found that the complexes are more active than their parent ligand.

4.15   SYNTHESIS OF MO (V) COMPLEXES USING EHMPB AND TRYPTAMINE

4.15.1   Synthesis of Ethyl Hydroxy Methyl Pyranyltryptamine Oxobutenoate Ligand (HL^1)

A mixture is made of tryptamine in ethanol and ethyl hydroxy methyl pyra-nyloxo-butenoate (**EHMPB**) in equal amounts. Then the reaction mixture is refluxed and kept in a cool place overnight. The resulting precipitate is yellow (HL^1) (**220**). By using the filtration method we can collect the compound. The purification of the compound will be done using the recrystallization method by using methanol.

4.15.2   Synthesis of Tryptaminoetylidene Methyl Pyrandione Ligand (HL^2)

The ligand HL^2 (**221**) (tryptaminoetylidene methyl pyrandione) can be synthe-sized by using details from previous literature [155].

220   **221**

4.15.3 Synthesis of Di-oxodimolybdenum Complex [Mo₂O₄(L¹)₂(CH₃OH)₂]

The metal complexes of molybdenum can be synthesized by suspending the complex [Mo₂O₃ (acac)₄] in dry methanol and adding the enaminone **HL¹** to it. Then heat the mixture up to 4 hours. Filter off the obtained crystalline product di-μ-oxo-bis (6-ethoxycarbonyl-4-oxo-2*H*-pyran-2-one)-3-[3-(tryptamino)-2-butenyl-1-onato-O,O¹]dimethanoldioxodimolybdenum   [Mo₂O₄(L¹)₂(CH₃OH)₂] (orange-red) (**222**) and wash the compound with cold alcohol and then dry [156].

222

4.15.4 Synthesis of Hexaoxohexamolybdate Complex (C₁₀H₁₂NH)[Mo₆O₁₂(OCH₃)₄(acac)₃]

The mixture of [Mo₂O₃(acac)₄] in dry methanol and tryptamine is heated gently. The crystalline product tryptaminium μ3-methoxo-tri-μ-methoxo-tri-μ3-oxo-tri-μ-oxo tris(acetylacetonato) hexaoxohexamolybdate (C₁₀H₁₂NH)[Mo₆O₁₂(OCH₃)₄(acac)₃] (orange-red) (**223**) will be formed from the remaining solution at room temperature in up to 5–10 days. Then filter off the product and wash with cold methanol.

223

Chapter 5

Various Biological Activities of DHA Derivatives

Archi Sharma

National Institute of Technology Raipur, Raipur, Chhattisgarh, India

5.1 INTRODUCTION

DHA is an important biologically active compound and it has antimicrobial activities [157]. It acts as a powerful antiseptic agent in an aqueous medium [158]. It is generally used in food industries, i.e., it can be used to protect vegetables by increasing the constancy of vitamin C during food processing [159].

It is also used as a preservative in fish sausages [160]. The metal complexes of DHA are promising fungicides, bactericides, herbicides, and insecticides [161,162].

Dehydroacetic acid (DHA) and its derivatives show various biological activities. A literature survey reveals that the reactions of DHA and its derivatives with different reagents can give versatile routes to the synthesis of a wide variety of biologically active compounds. The biological activities of DHA and its derivatives are shown in Fig. 5.1.

5.2 ANTI-HIV ACTIVITY

The human immunodeficiency virus [163] (HIV) is a lentivirus (Fig. 5.2), one of the retrovirus subgroups, which causes the viral infection (HIV) and syndrome, namely acquired immunodeficiency syndrome (AIDS). In a progressive manner, the failure of the immune system takes place in the condition of AIDS and allows all types of life-threatening infections to thrive. Depending on the HIV subtype, the average survival time of a person after being infected by the virus without any treatment is estimated to be 9–11 years.

HIV infects vital cells such as helper T-cells (specifically CD^{4+} T-cells), macrophages, and dendritic cells, which are essential in maintaining the human immune system. HIV infection is responsible for declining the levels of CD^{4+} T-cells and apoptosis of uninfected cells. When the CD4+ T-cell count decreases below the extreme level, then the body becomes progressively more prone to various infections.

Dehydroacetic Acid and Its Derivatives. DOI: http://dx.doi.org/10.1016/B978-0-08-101926-9.00005-5
© 2017 Elsevier Ltd. All rights reserved.

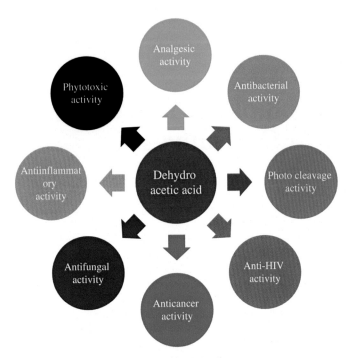

FIGURE 5.1 Biological activities of DHA and its derivatives.

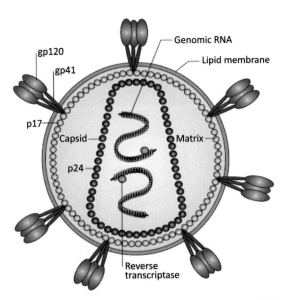

FIGURE 5.2 Structure of HIV. *Gunilla BKH, Ron AMF, Sanjay P, Dennis RB, Joseph S, Richard TW. The challenges of eliciting neutralizing antibodies to HIV-1 and to influenza virus. Nat Rev 2008;6:143–55.*

5.2.1   Antiviral Drugs

In 1987, major advances in antiretroviral therapy were made by the introduction of zidovudine (AZT). With the help of highly active antiretroviral therapy (HAART), HIV-1 infection can be manageable in patients who have access to medication and virologic suppression. About 20 drugs have been currently approved for HIV-1 clinical therapy.

5.2.2   3-Acyl-4-Hydroxypyrones and Their Boron Difluoride Complexes as HIV-1 Integrase Inhibitors

In 2007, USFDA (United States Food and Drug Administration) approved the first class drugs, namely integrase inhibitors. HIV produces an enzyme called integrase which enables its DNA to integrate or merge with DNA present in the cells of an infected person (Fig. 5.3A). The virus is incapable of reproducing on its own and spreads by the reproduction of the host cells. Integrase inhibitor drugs prevent the integration of HIV genetic material with human DNA material [164,165] (Fig. 5.3B).

DHA and its boron difluoride complexes (**225**) exhibit anti-HIV activity. The cinnamoyl pyrones (**224**) can be used as HIV-1 integrase inhibitors due to their antiviral activity with HIV infected cells which has been convincingly confirmed by in vitro experiments of the compound [166].

| 224 | 225 |

It is found that active complex, namely the difluoro complex of DHA, plays a crucial role in inhibition by forming strong interactions with the amino acid residues D64, E152, and Mg^{2+} ions. The binding of the difluoro compound with integrase enzyme is shown in Fig. 5.4.

The antiviral activity of a unique class of integrase enzyme inhibitors, namely DHA derivatives, was revealed by both computer-assisted screening and in vitro tests.

5.3   ANTICANCER ACTIVITY

Cancer is a disease which is described by uncontrolled growth and development of cell. There are more than one hundred varieties of cancer and each cancer is categorized by the kind of cell that is first affected. In the year 2012, the World

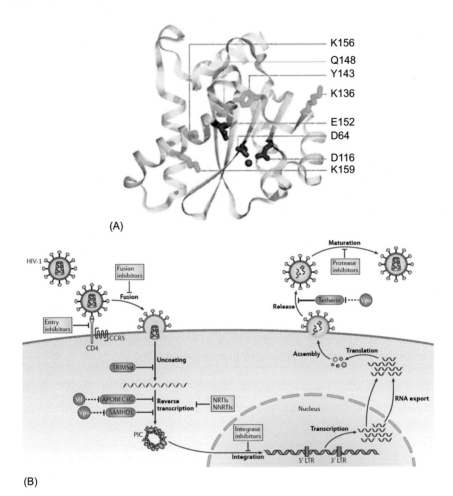

FIGURE 5.3   (A) Structure of HIV-1 integrase. (B) Schematic overview of the HIV-1 replication cycle. *Yves P, Allison AJ, Christophe M. Integrase inhibitors to treat HIV/AIDS. Nat Rev 2005;4:236–48 OR Françoise B-S, Anna LR, Jean-François D. Past, present and future: 30 years of HIV research, Nat Rev 2013;11:877–83.*

Health Organization (WHO) estimated that there were about 4 million cancer cases and 8.2 million cancer linked deaths worldwide.

In cancer, the uncontrollable division of altered cells leads to the formation of lumps/bulgings or masses of tissue known as tumors [167] (except in the case of leukemia) (Fig. 5.5). Tumors can show their negative impact on the digestive, circulatory, and nervous systems by releasing their respective chemical messengers into the body. Tumors that stay in a particular area with moderate growth are usually considered as benign. When a tumor spreads from one particular

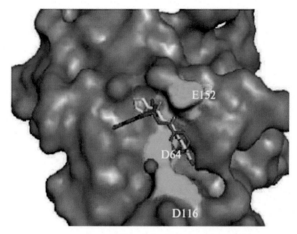

FIGURE 5.4 Integrase with difluoro compound. *Tambov KV, Voevodina IV, Manaev AV, Ivanenkov YA, Neamati N, Travena VF. Structures and biological activity of cinnamoyl derivatives of coumarins and dehydroacetic acid and their boron difluoride complexes, Russ Chem Bull 2012;61:78–90.*

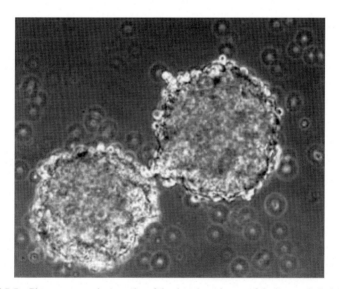

FIGURE 5.5 Phase-contrast photography of the ductal carcinoma of the breast. *Ja-Lok K, Sung-Chan P, Kyung-Hee K, You-Kyung J, Sung-Hee K, Young-Kyoung S, et al. Establishment and characterization of seven human breast cancer cell lines including two triple-negative cell lines. Int J Oncol 2013;43:2073–81.*

area to different areas of the body, where it grows and causes harm to other living tissues, then it is known as metastasis which is very difficult to treat.

5.3.1 DHA-Hydrazone and its Zn (II), Ru (III), and Pd (II) Metal Complexes as Anticancer Agents

DHA-hydrazone and its Ru (III), Pd (II), and Zn (II) metal complexes can be used as anticancer agents [168]. The condensation of **DHA** with ethylene diamine (en) leads to the formation of Schiff base ethylenediamine bis-DHA hydrazone (H_2L) (**226**).

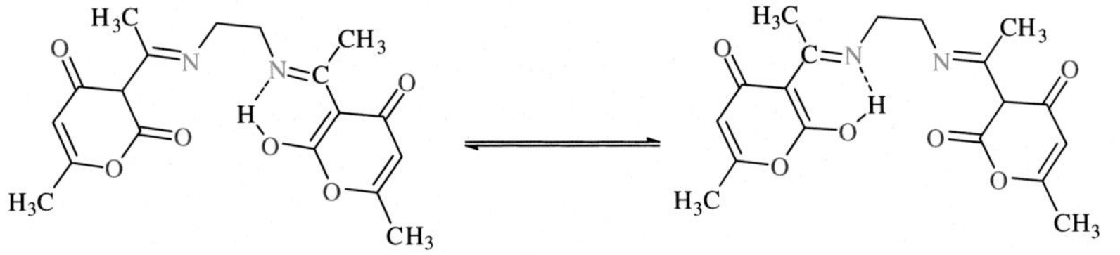

226

5.4 ANTILIVER CANCER ACTIVITY

In the human body the liver can also be affected by cancer. Cancer affected liver [169] is shown in Fig. 5.6A. The cytotoxicity of the ligand (H_2L) and its Ru (III), Pd (II), and Zn (II) metal complexes on human liver carcinoma cells lines **HepG2** (Fig. 5.6B) [170] were determined.

It was observed that the Pd (II) complex of ethylenediamine bis-DHA hydrazone ligand is more active against **HepG2** cells than the free ligand (H_2L) and shows more cytotoxic activity with a lower IC_{50} value of $2.67 \mu g/mL$ as compared to other complexes. It indicates that the Pd (II) complex (**227**) is more active than the reference drug doxorubicin (**228**) ($IC_{50} = 3.73 \mu g/mL$).

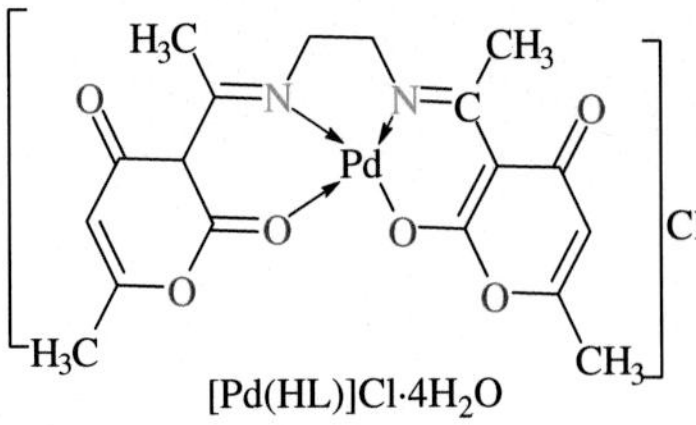

[Pd(HL)]Cl·4H₂O

227 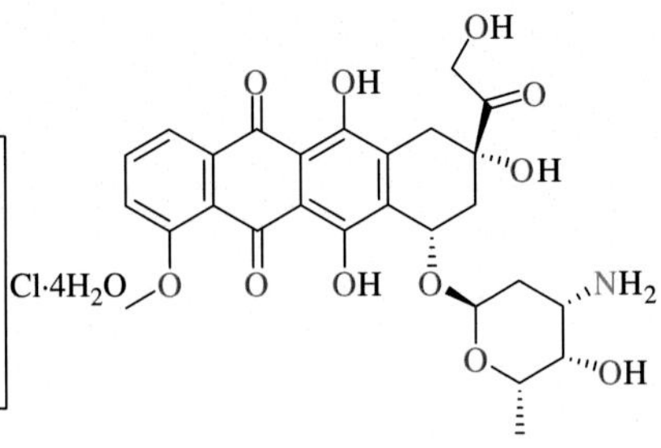   **228**

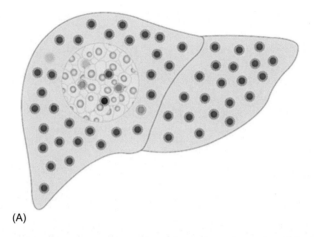

(A)

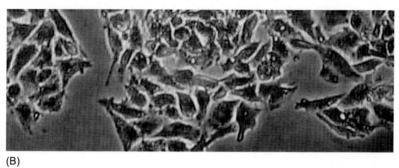

(B)

FIGURE 5.6 (A) Hepatocellular carcinoma. (B) HepG2 cells. *(A) Holmes JA, Chung RT. HCV compartmentalization in HCC: driver, passenger or both? Nat Rev 2016. doi:10.1038/nrgastro.2016.46. (B) Sakineh KN, Michael W. Transcriptional down regulation of hTERT and senescence induction in HepG2 cells by chelidonine, World J Gastroenterol 2009;15:3603–10.*

5.4.1 Antibreast Cancer Activity

The cytotoxic activity of the ligand (H_2L) and its Zn (II), Ru (III), and Pd (II) complexes on breast carcinoma cells lines **MCF-7** have been determined. It is found that the DHA–hydrazone ligand (H_2L) and its Zn (II) and Ru (III) metal complexes showed similar effects on **MCF-7** [171] breast cancer cells (Fig. 5.7).

The Pd (II) complex has more cytotoxicity with an IC_{50} value of 3.28 µg/mL which is nearly equal to the IC_{50} of the reference drug doxorubicin ($IC_{50} = 2.97$ µg/mL).

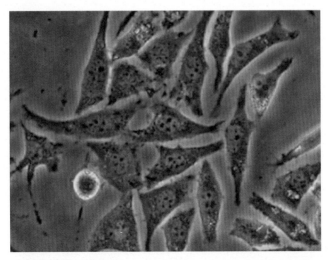

FIGURE 5.7 MCF-7 human cell line. *Chaoran L, Zhong L, Meng L, Xiaoling L, Yum-Shing W, Sai-Ming N, Wenjie Z, Yibo Z, Tianfeng C. Enhancement of auranofin-induced apoptosis in MCF-7 human breast cells by selenocystine, a synergistic inhibitor of thioredoxin reductase. PLoS One 2013;8:1–14.*

[Zn(HL)(H₂O)₂]Cl·4H₂O (229)

[Ru(HL)Cl₂]2H₂O (230)

[Pd(HL)]Cl·4H₂O (231)

5.5 ANTIBACTERIAL ACTIVITY

5.5.1 Different Heterocyclics From DHA Enaminone

The antibacterial activity of **DHA** derivatives (Scheme 5.1, **234–241**), such as a large variety of heterocyclics from DHA enaminone, was analyzed [172].

The in vitro antibacterial activity of the **DHA** derivatives was evaluated against both gram-positive bacteria such as *Bacillus subtilis* [173] (Fig. 5.8A) and *Bacillus thuringiensis* [174] (Fig. 5.8B) and gram-negative bacteria such as *Pseudomonas aeruginosa* and *Escherichia coli*. Chloramphenicol (**232**), cephalothin (**233**), and cycloheximide can be used as reference or standard drugs to test their antimicrobial activities.

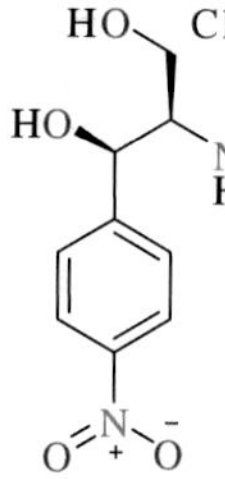

232 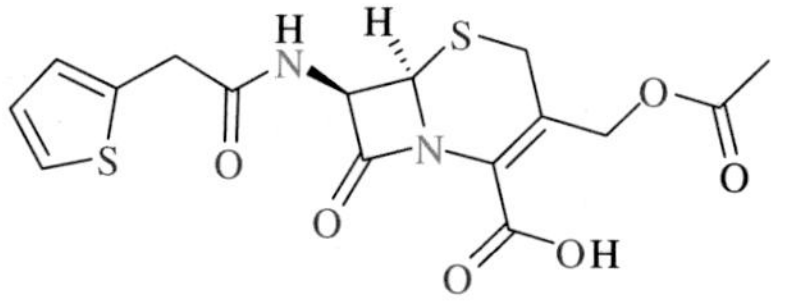 **233**

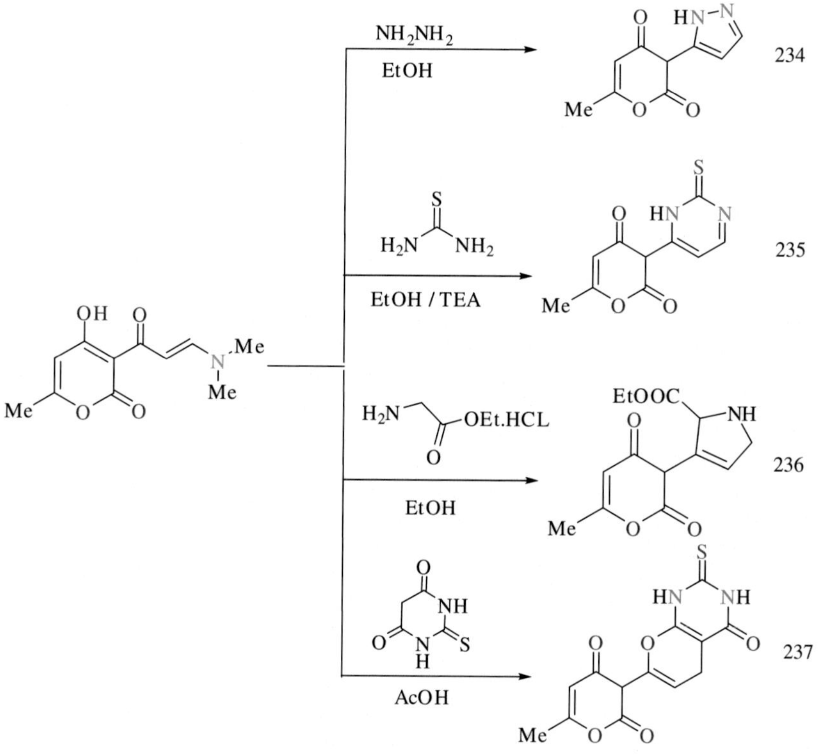

SCHEME 5.1 Synthesis of DHA derivatives.

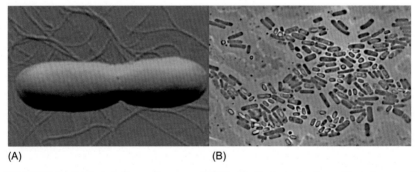

SCHEME 5.1 (Continued)

FIGURE 5.8 (A) Atomic force microscopy (AFM). (B) *B. thuringiensis* amplitude images of *B. subtilis*. (A) Shaobin L, Andrew KN, Rong X, Jun W, Cher MT, Yanhui Y, Yuan C. Antibacterial action of dispersed single-walled carbon nanotubes on Escherichia coli and Bacillus subtilis investigated by atomic force microscopy. Nanoscale 2010;2:2744–50. (B) Mario R-L, Montserrat R-S. Biological control of mosquito larvae by Bacillus thuringiensis subsp. israelensis, insecticides— pest engineering. In: Farzana P, editor. In tech; 2012. p. 239–64 [Chapter 11].

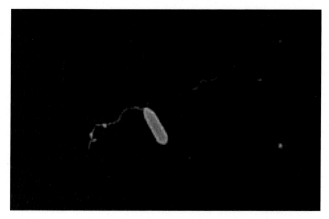

FIGURE 5.9 A high-power scanning electron micrograph showing the adherence of *P. aerugi-nosa* bacilli to the collagen surface. *Tsang KW, Shum DK, Chan S, Ng P, Mak J, Leung R, et al. Pseudomonas aeruginosa adherence to human basement membrane collagen in vitro. Eur Respir J 2003;21:932–8.*

The ability of compounds **234, 235, 237**, and **238** in inhibiting the growth of *B. subtilis* was equal to chloramphenicol (MIC 3.125 μg/mL) However, in the case of *B. thuringiensis*, the antibacterial activity of compounds **234, 235**, and **238** was 50% lower than that of chloramphenicol and also at the same time, the compound **237** exhibited equal potency to chloramphenicol. The compounds **236, 239, 240**, and **241** exhibited equal potency to cephalothin in inhibiting the growth of *B. subtilis* and *B. thurigiensis*, whereas they were 50% lower than that of chloramphenicol (MIC 6.25 μg/mL) (MIC 6.25 μg/mL). On the other hand, the compounds **234** and **235** showed equal antibacterial activity to chloramphenicol and cephalothin against *P. aeruginosa* [175] (Fig. 5.9) and *E. coli*.

5.5.2 4-Hydroxy-3-[(2*E*)-3-(4-Hydroxyphenyl) Prop-2-Enoyl]-6-Methyl-2*H*-Pyran-2-One (Ligand) and its Various Metal Complexes

The Mn (II), Fe (III), Co (II), Ni (II), and Cu (II) metal complexes with ligand (**243**) were synthesized from **DHA** and 4-hydroxy benzaldehyde (**242**).

242

The in vitro antibacterial action of the above synthesized ligand (**242**) and its various metal complexes were examined against both gram-negative, such as

E. coli, and gram-positive, such as *Staphylococcus aureus*, bacterial strains by utilizing neomycin (**244**) as a reference or standard antibiotic [176].

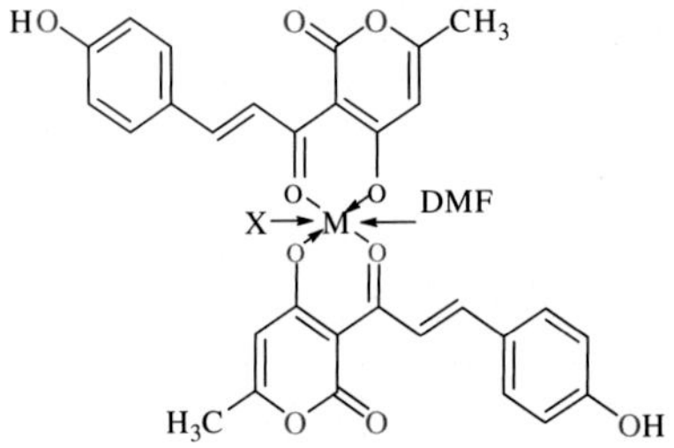

243 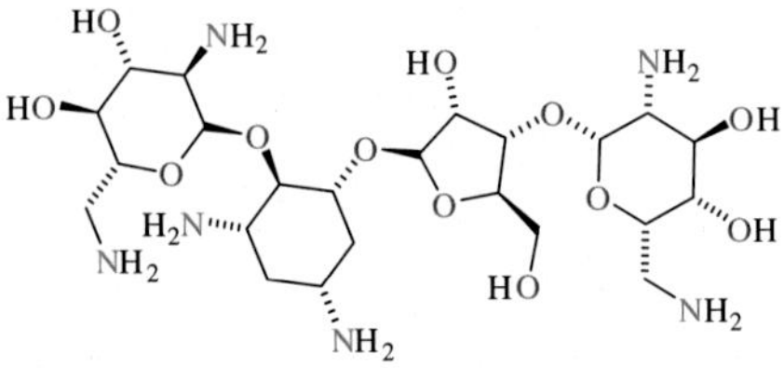   **244**

   The results showed that the ligand exhibited weak antibacterial activity, whereas its complexes showed moderate activity against *S. aureus* and *E. coli*. Due to the chelating property of the ligand in the metal complex, it can act as a more powerful and potent bactericidal agent (killing the bacteria) than the non-chelated ligand. Actually, chelation reduces the metal ion's polarity in a complex because of delocalization of the π electrons over whole chelate ring and partial sharing of its positive charge with donor groups.

5.5.3   Ruthenium Complexes of DHA

The antibacterial activity of **DHA** complexes of Ru (II) and Ru (III) with $PPh_3/AsPh_3$ (**245–249**) was evaluated [140]. The biological activity of a large number of Ru (II) and Ru (III) metal complexes has been explored. The **DHA** (ligand) and its Ru (II) and Ru (III) metal complexes were examined against five disease-causing bacteria.

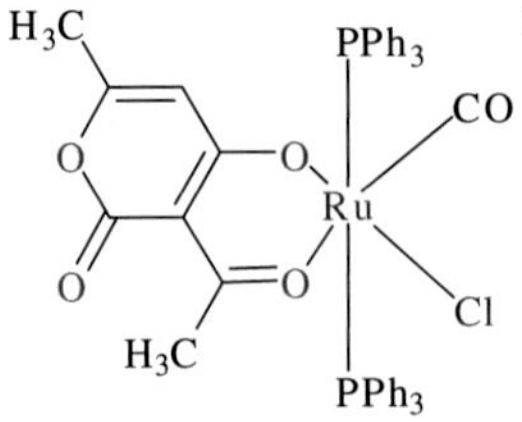

245 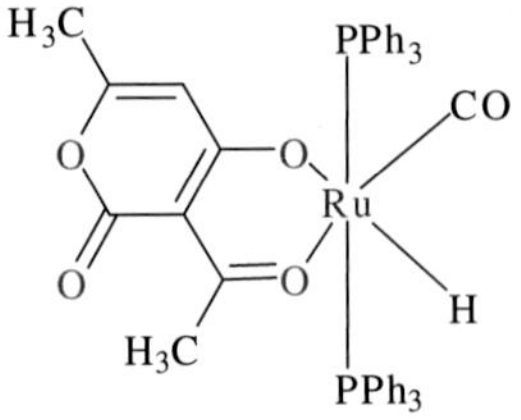   **246**

247   248

249

The **DHA (ligand)** and its Ru (II) and Ru (III) metal complexes were observed to retain antibacterial action against both gram-positive, such as *S. aureus* and *Staphylococcus epidermidis*, and gram-negative, such as *Shigella sonnei*, *E. coli*, and *Klebsiella pneumoniae*, bacterial strains. The antibacterial activity of all these complexes depends upon the type of bacteria. For instance, the complex **247** shows good activity against gram-negative bacteria *S. sonnei*, complexes **245** and **249** shows activity against gram-negative bacteria *K. pneumonia*, and complexes **248** and **246** show activity against gram-positive bacteria *S. aureus*.

5.5.4 DHA Thiosemicarbazone and its Ruthenium Complexes

The DHA thiosemicarbazone (**250**) and its ruthenium complexes (**251**, **252**) have been synthesized and their antibacterial activities explored by using ciprofloxin (5 µg) as a reference or standard drug. A series of ruthenium (II)-containing DHA thiosemicarbazone complexes, namely [Ru(dhatsc)(CO)(B) (EPh$_3$)] (where E = P, B = PPh$_3$, py, pip, or E = As, B = AsPh$_3$; dhatsc = dibasic tridentate DHA thiosemicarbazone), was synthesized by refluxing [RuHCl(CO) (B)(EPh$_3$)$_2$] (where E = As/P, B = AsPh$_3$/PPh$_3$, pip, py) and DHA thiosemi-carbazone (abbreviated as H$_2$dhatsc, where H$_2$ means two dissociable protons) in the presence of benzene. These ruthenium (II) complexes of **DHA** exhibited moderate activity in preventing the growth of both gram-negative bacteria (*E. coli*) and gram-positive bacteria (*S. aureus*) [177].

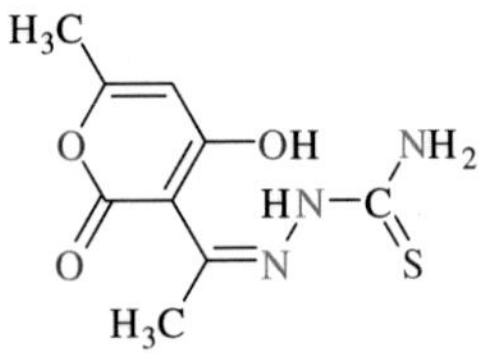

250

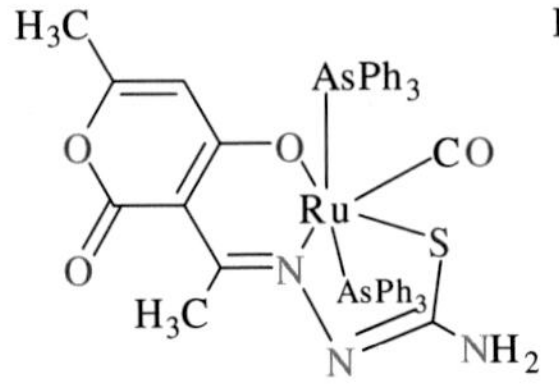

251 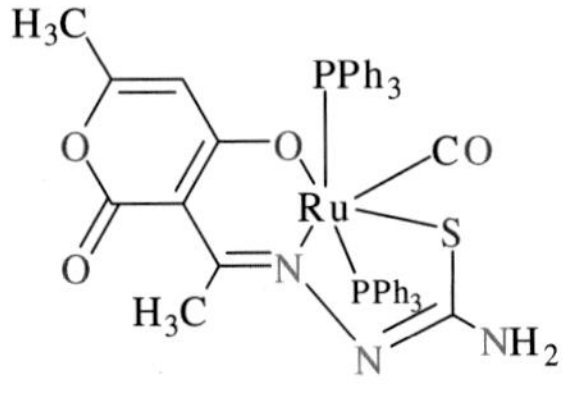   **252**

5.5.5 4-Hydroxy-3(1-{2-(Benzylideneamino)-Phenylimino}-Ethyl)-6-Methyl-2*H*-Pyran-2-One and its Complexes

The Cu (II), Mn (II), Co (II), Ni (II), and Fe (III) metal complexes with the chlorobenzylidene aminophenylimino derivative of DHA **(tridentate Schiff base ligand) (253)** were synthesized from DHA, *o*-phenylenediamine, and *p*-chloro benzaldehyde [149].

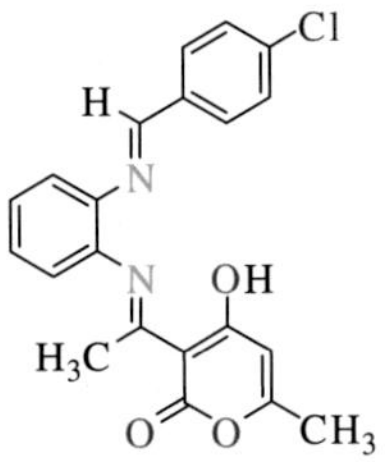

253

The biological activity, such as the antibacterial activity, of copper complexes of DHA is greater than the copper salts, DHA, and its other complexes. The antibacterial activity of ligand **(253)** and its various metal complexes against the bacterial species *S. aureus* and *E. coli* were tested by using the paper disk technique and the results obtained were compared with a standard or reference drug, namely ciprofloxin. The metal chelates show their antibacterial activity in the following sequence: **Cu > Mn> Fe > Co > Ni**, which does not depend on the stability order of metal ions.

5.5.6 Dichlorophenylimino Derivative of DHA

The antibacterial activity of the ligand namely dichlorophenyliminoethyl-hydroxy-methyl-pyranone (**254**) was screened against both gram-positive (single-layered), such as *S. aureus* and *B. subtilis*, and gram-negative (multilayered and pathogenic), such as *E. coli* and *P. aeruginosa*, bacterial species. Imipenem (**255**) was used as a reference or standard drug under similar experimental conditions, which has showed 100% ability in retarding the cultured bacterial growth [178].

254   255

It showed an enhanced antibacterial property against *S. aureus*, *B. subtilis*, *E. coli*, and *P. aeruginosa* bacterial species.

5.5.7 Chloro-Hydroxy-Methyl-Oxo-Pyranylethylideneacetohydrazide and its Metal Complexes

The in vitro antibacterial activity of the compound chloro-hydroxy-methyl-oxo-pyranylethylideneacetohydrazide (a chloro derivative of DHA hydrazone) and its corresponding transition metal complexes, such as **Co (II), Ni (II), Cu (II), and Zn (II)** (**256**), were screened against both gram-positive bacteria, such as *B. subtilis* and *Micrococcus luteus*, and gram-negative bacteria, such as *P. aeruginosa* and *Pseudomonas mendocina*, by using the agar plate disk method and streptomycin (**257**) as a reference or standard drug [179]. The antibacterial activity of the copper (II) complex was highest against gram-positive bacterial species even at a lower concentration of 25 μg/mL, which is very close to streptomycin. The structure of DHA hydrazone and its metal complexes are given as follows.

256   257

5.5.8 3-Cinnamoyl Derivatives of DHA

The antibacterial activity of cinnamoyl derivatives of DHA (**258**, **259**, **260**, and **261**) was screened against gram-positive (*S. aureus* and *B. subtilis*), gram-negative (*Salmonella typhi* and *E. coli*), and bacterial species by using penicillin as a reference or standard drug. The compounds **258** and **260** showed excellent antibacterial activity against all the three bacterial species except *S. typhi*. The inhibition zone (14 mm) shown by compound **258** against *E. coli* was higher than the reference drug penicillin (13 mm), which clearly indicates its strong antibacterial activity [180].

258   **259**

260   **261**

5.5.9 Hydroxynaphthalenylmethylene Aminophenyliminoethyl Derivative of DHA and its Metal Complexes

The in vitro antibacterial activity of 4-hydroxy-3-((*E*)-1-(2-((*E*)-(2-hydroxynaphthalen-1-yl) methyleneamino)phenylimino)ethyl)-6-methyl-2*H*-pyran-2-one (**ligand**) and its Cu (II), Mn (II), Ni (II), Co (II), and Fe (III) metal complexes (**262**) were assessed against the bacterial species *E. coli* and *S. aureus* using the paper disk plate technique and ciprofloxacin as a reference antibiotic. The activity of the complexes was tested at the concentration of 500 and 1000 ppm and compared with ciprofloxacin [181] (**263**). When compared with a free ligand, these complexes are biologically active and showed improved antibacterial activity.

R = H/CH₃

263

M = Ni (II), Mn (II), Mn (II),
Fe (III), and Cu (II)

262

5.5.10   Hydroxynaphthalenylmethylene Aminophenyliminoethyl Derivative of DHA and its Lanthanide Metal Complexes

The 4-hydroxy-3-(((E)-1-(2-((E)-(2-hydroxynaphthalen-1-yl) methyleneamino) phenylimino)ethyl)-6-methyl-2H-pyran-2-one and its lanthanide metal complexes (**264**) showed more enhanced antibacterial activity than the free ligand. The antibacterial property of the ligand and its lanthanide metal complexes were examined against the bacterial species *E. coli* and *S. aureus* using the paper disk plate technique and ciprofloxin as a reference drug [182]. The results indicate the tremendous rise in the activity of metal complexes from La (III) to Gd (III) which is due to the increasing size of the metal ions, the increasing order of stability constants, and the increasing trend in magnetic moments.

M = Gd (III), Nd (III), Pr (III), La (III), Sm (III), and Ce (III)

264

5.5.11   DHA Chalcones and Pyrazolines

The antibacterial activity of DHA chalcones (**265** and **266**) and pyrazolines (**267**)a, (**267**)b, (**268**)a, and (**268**)b was assessed by using chloramphenicol as a reference or standard drug. The activity of chalcone (**265**) was examined against both gram-positive (*B. subtilis*) and gram-negative (*E. coli*) bacterial species. The inhibition zone (20 mm) of compound **265** was higher than the inhibition zone (21 mm) of chloramphenicol which indicates a powerful antibacterial activity. Among chalcones, the compound (**265**) exhibited good activity against both bacterial species *S. aureus* and *Salmonella typhimurium*, whereas the other compound **266** showed moderate activity against both types of bacteria. Among pyrazolines, the compound **267b** showed a high degree of inhibition against *S. typhimurium* (gram-negative bacteria) and moderate inhibition against *E. coli* (gram-negative bacteria) whereas it did not show any activity against any type of gram-positive bacteria. The other pyrazolines **267a** and **268a** exhibited moderate to good activity against both gram-positive and gram–negative bacteria as compared to chloramphenicol [183]. The compound **268b** also showed moderate activity against gram-negative bacteria and no activity against gram-positive bacteria.

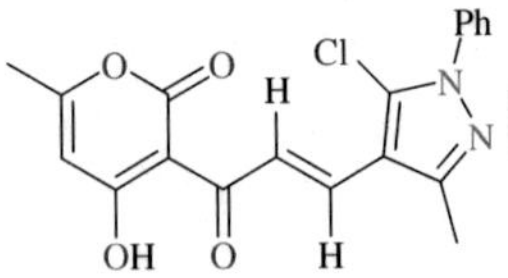

265 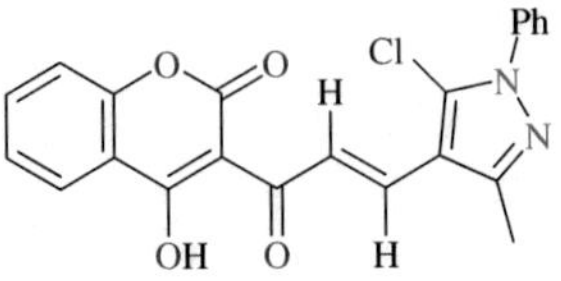   266

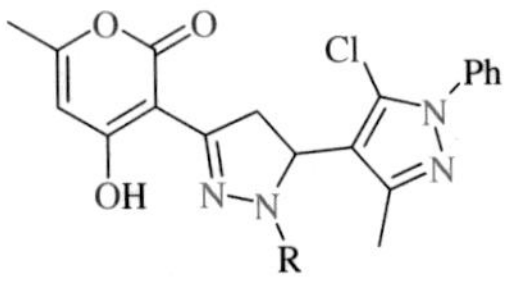

267a = R = H,   268c = R = H,
267b = R = Ph   268d = R = Ph

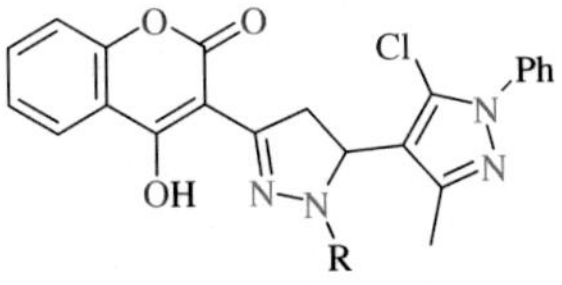

5.5.12 4-Hydroxy-3-((*E*)-1-(4-((*Z*)-1-(2-Hydroxyphenyl) Ethylideneamino)-6-Methyl-1,3,5-Triazin-2-Ylimino)Ethyl)-6-Methyl-2*H*-Pyran-2-One and its Metal Complexes

The antibacterial activity of 4-hydroxy-3-((*E*)-1-(4-((*Z*)-1-(2-hydroxyphenyl) ethylideneamino)-6-methyl-1,3,5-triazin-2-ylimino)ethyl)-6-methyl-2*H*-pyran-2-one (ligand) (**269**) and its various metal complexes (**270**) were assessed against the bacterial species *E. coli*, *B. subtilis*, *S. aureus*, and *K. pneumoniae* using the paper disk plate method and the reference or standard drug tetracycline (**271**) [184,185].

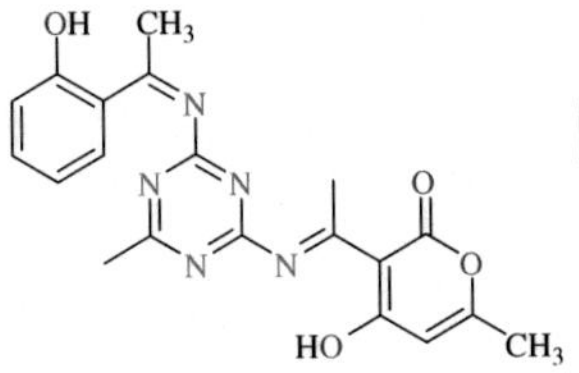

269   270

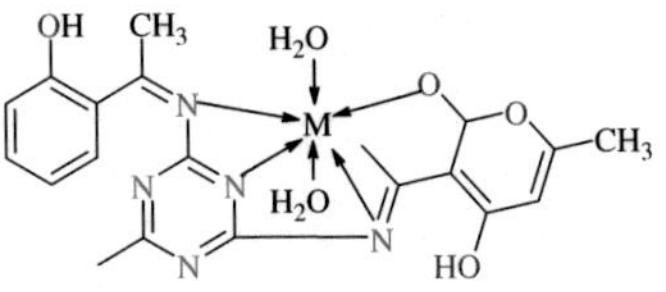

M=Cr (III), Fe (III), and Co (II)

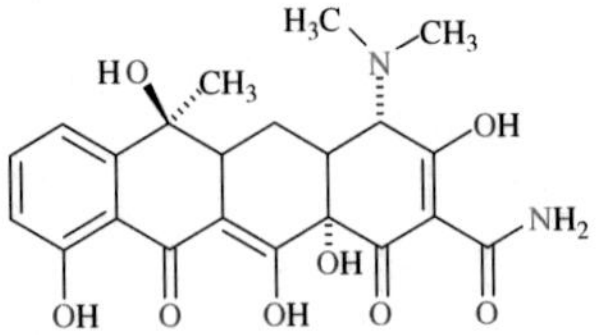

271

5.5.13 Pyrimidine Derivatives

The pyrimidine derivatives fluorophenyl-hydroxy-6-methyl-2-oxo-pyranyl-benzylthio pyrimidine (**272**), thiophene-hydroxy-methyl-oxo-pyranyl-ben-zylthiopyrimidine (**273**), and pyridyl-hydroxy-methyl-oxo-pyranyl-benzylthio pyrimidine (**274**) showed mild activity against the gram-positive bacteria *B. subtilis*. The thiophene-hydroxy-methyl-oxo-pyranyl-benzylthiopyrimidined-erivative (**273**) also showed mild activity against the gram–negative bacteria *P. aeruginosa*. The antibacterial activity of the above pyrimidine derivatives was checked by using ciprofloxacin as a standard or reference drug [186].

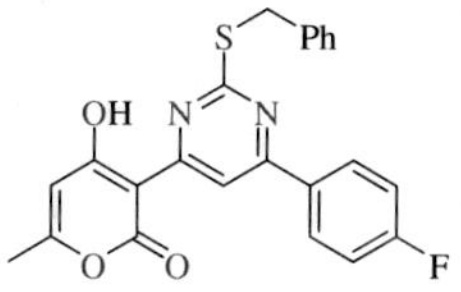

272 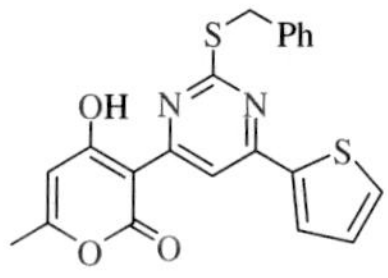   273

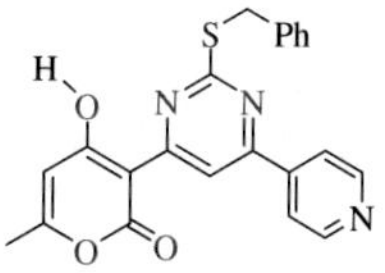

274

5.5.14 2-Aminoethyl Ethane-1,2-Diamine Derivative of DHA and its Metal Complexes

The antibacterial activity of the 2-aminoethyl ethane-1,2-diamine derivative of DHA (ligand) and its **Ni (II), Cu (II),** and **Mn (II)** metal complexes (**275**) were assessed against both the gram-negative bacteria *E. coli* and the gram-positive bacteria *S. aureus*. These metal complexes show slight variations in the degree of inhibition on the growth of the bacteria [187].

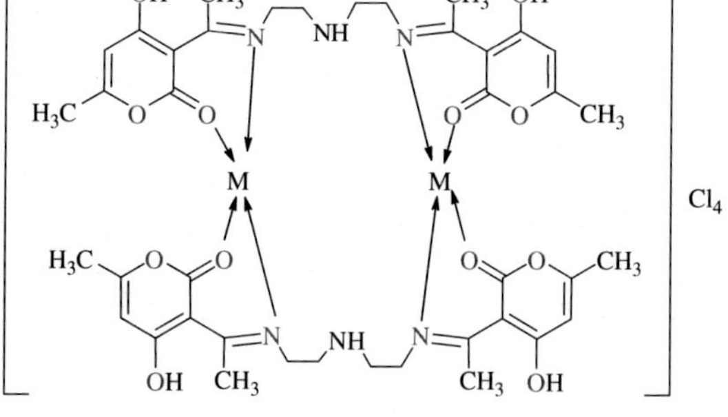

M = Ni (II), Cu (II), and Mn (II)

275

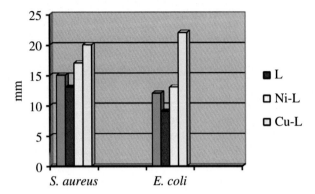

GRAPH 5.1 The effect of ligands and their metal complexes toward the bacteria.

The impact of the ligand and its various metal complexes toward the bacterial growth are shown in Graph 5.1.

5.5.15 Cyano Ethylidene Acetohydrazide Derivative of DHA and its Metal Complexes

The in vitro antibacterial activity of the cyano ethylidene acetohydrazide derivative of DHA (**ligand**) and its metal complexes (Zn (II), Cu (II), Ni (II), Mn (II), and Co (II)) (**276**) were screened against two gram-positive bacteria, *S. aureus* and *B. subtilis*, and two gram-negative bacteria, *Pseudomonas syringae* and *P. aeruginosa*, and the activity was compared by taking oxacillin (**277**) as a reference or standard antibiotic [188].

276   **277**

The metal complexes showed higher in vitro antibacterial activity than the free ligand under similar experimental conditions. Chiefly, the copper complex showed better antibacterial activity than the other metal complexes. The antibacterial activity of the ligand can be enhanced by the slight modifications in its structure.

5.5.16 Dichlorophenylimino Ethyl Derivative of DHA

The Schiff base dichlorophenylimino ethyl derivative of DHA (**278**) showed better antibacterial activity against *B. subtilis*, *S. aureus*, *E. coli*, and *P. aeruginosa*. The imipenem was taken as a reference antibiotic for assessing the antibacterial activity of Schiff bases [178].

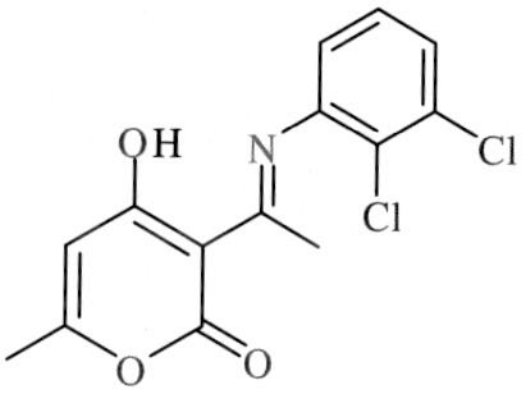

278

5.5.17 Dihydroxyphenyl Derivative of DHA

The antibacterial activity of dihydroxyphenyl derivative of DHA (**279**) was screened against two gram-negative bacteria, *S. typhi* and *E. coli*, and two gram-positive bacteria, *B. subtilis* and *S. aureus*, by following the agar cup method. The compound (**279**) was found to be a good antibacterial agent against all types of bacteria [180].

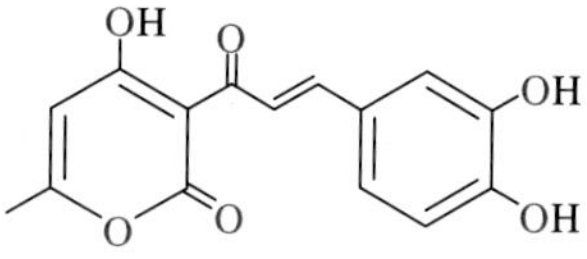

279

5.6 ANTIFUNGAL ACTIVITY

A fungus is a eukaryotic organism and includes both unicellular microorganisms, such as yeasts and molds, as well as multicellular fungi that produce

FIGURE 5.10 *Rhizopus stolonifer* growing on blackberry fruits. *Ismael FC-D, Valentina A-P, Sigifredo L-D, Miguel GV-DV, Ana NH-L. Antagonistic bacteria with potential for biocontrol on Rhizopus stolonifer obtained from blackberry fruits. Fruits 2014;69:41–6.*

familiar fruiting forms known as mushrooms. The growth of *Rhizopus stolonifer* [189] on blackberry fruits is shown in Fig. 5.10.

DHA and its derivatives, Schiff bases, and metal complexes can act as good antifungal agents. Some of them are described below.

5.6.1   3-/2-Aminophenylamine-(*p*-Toluoyl)-4-Hydroxy-6-(*p*-Tolyl)-2*H*-Pyrane-2-One and 4-Hydroxy-3-(*p*-Toluoyl)-6-(*p*-Tolil)-2*H*-Pyrane-2-One

The antifungal activity of DHA analogs 3-/2-aminophenylamine-(*p*-toluoyl)-4-hydroxy-6-(*p*-tolyl)-2*H*-pyrane-2-one (Schiff base) (**280**) and 4-hydroxy-3-(*p*-toluoyl)-6-(*p*-tolil)-2*H*-pyrane-2-one (**DHT**) (**281**) inhibit the growth of mycotoxin-producing molds and the accumulation of aflatoxin B_1 (AFB_1) and ochratoxin A (OTA). They also effectively inhibit the accumulation of vomitoxin (**282**) (deoxynivalenol) (DON) by *Fusarium graminearum* [190]. *Fusarium*-affected wheat and maize [191] are shown in Fig. 5.11.

The effect of the DHA analogs in controlling the growth and accumulation of deoxynivalenolhas been studied using a mold *F. graminearum* ZMPBF 1244 and a maize grain hybrid. The inhibitory effect of DHA analogs is considered to be due to the inhibition of growth rather than toxin accumulation.

FIGURE 5.11   Infection of *F. graminearum* on (A) wheat and (B) maize. *Christian AV, Wilhelm S, Siegfried S. A secreted lipase of Fusarium graminearum is a virulence factor required for infection of cereals. Plant J 2005;42:364–75.*

280   **281**

The antifungal activity of Schiff base and DHT was analyzed by using two techniques, i.e., standard assay and assay in flask culture.

L^1	$4-H_3C-C_6H_4$
L^2	$4-H_3CO-C_6H_4$
L^3	$4-Cl-C_6H_4$
L^4	$3, 4(H_3CO)_2C_6H_3$

282

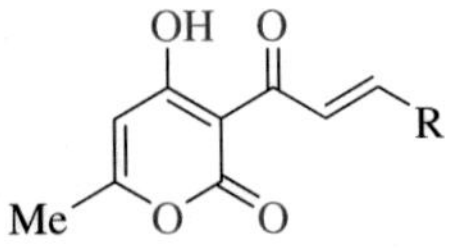

FIGURE 5.12   General structure of the chalcone ligands. *Christiane H-F, Arnab P. Specialist fungi, versatile genomes. Nat Rev 2007;5:332–3. (B) Walther G, Pawłowska J, Alastruey-Izquierdo A, Wrzosek M, Rodriguez-Tudela JL, Dolatabadi S, et al. DNA barcoding in Mucorales: an inventory of biodiversity. Persoonia 2013;30:11–47.*

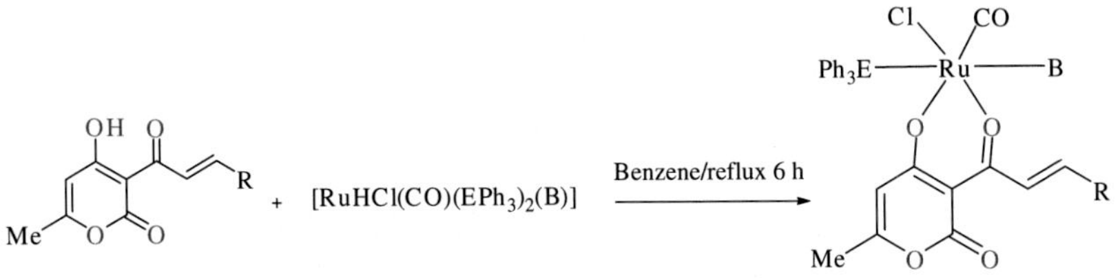

SCHEME 5.2   Synthesis of Ru metal complexes from DHA chalcones.

The newly synthesized analogs of DHA (Schiff base and DHT) control the growth of toxigenous mold *F. graminearum* ZMPBF 1244 and the accumulation of vomitoxin (DON) in maize grain (FAO 280 Os, 298 P hybrid) in certain parameters of cultivation. By using different concentrations of Schiff base and DHT, we found that the Schiff base was the better analog for the inhibition of mold growth and vomitoxin accumulation.

5.6.2   Ruthenium Chalconate Complexes

Chalcone ligands (**283**) which are obtained from DHA and aldehydes are used to synthesize new ruthenium (II) chalconate complexes. The general structure of the chalcone ligands is shown in Fig. 5.12.

By using the above chalcones as one of the reactants, the ruthenium complexes were synthesized. In this, chalcones (**283**) are refluxed with a ruthenium metal complex (**284**) in the presence of benzene to produce chalcone ruthenium metal complexes (**285**) (Scheme 5.2).

E = P or As
B = PPh$_3$, AsPh$_3$, or Py
R = 4-H$_3$C-C$_6$H$_4$, 4-H$_3$CO-C$_6$H$_4$, 4-Cl-C$_6$H$_4$, or 3,4-(H$_3$CO)$_2$C$_6$H$_3$

The antifungal activity of ligands and their ruthenium complexes was verified against the pathogenic fungi *Aspergillus niger* [192] (Fig. 5.13A) and *Mucor* species [193] (Fig. 5.13B). The fungi were cultured on dextrose agar medium of sabouraud, incubated at 30°C, and used as the test, whereas dimethyl sulfoxidesolvent was used as the control.

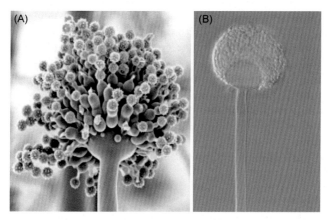

FIGURE 5.13   (A) SEM showing asexual sporulation in *A. niger*. (B) Structure of *Mucor luteus*. *Zadah ME, Jafari AA, Sedighi S, Seifati SM. Int Jour Pharm Ther 2016;7:90–96.*

The inhibition of fungal growth can be expressed in terms of percentage and determined by using the following equation:

$$\text{Inhibition} \% = 100(C - T)/C$$

where

C = diameter of fungal growth on the control plate;
T = diameter of fungal growth on the test plate.

The in vitro antifungal screening of the ligands and their metal complexes was assessed by using disk diffusion procedure against *A. niger* and *Mucor* species. The complexes of Ru (II) were more toxic than their parent ligands against the same fungal species under equal experimental conditions. The impact of the metal ion on the normal cell process is responsible for improving the antifungal activity of metal chelates.

5.6.3   DHA and Ozonized Water on Pistachio Nuts

Aflatoxin is the most powerful group of carcinogenic mycotoxins produced mainly by *Aspergillus flavus* on pistachio nuts.

DHA and ozonized water can be used to control the growth of *A. flavus* and the accumulation of aflatoxin B_1 (**286**) on pistachios. The inhibitory effect of DHA on the growth of fungi and the production of aflatoxin was much more than ozonized water.

DHA and ozonized water are very efficient antimycotoxigenic and antifungal agents. The application of different concentrations of **DHA** showed more antifungal activity in controlling the growth of *A. flavus* and the accumulation of aflatoxinin compared with ozonized water for the treatment of pistachios nuts [194].

5.6.4 Hydroxyphenylpropenoylmethylpyran-2-One and its Copper Complex

The 4-hydroxy-3-[(2*E*)-3-(4-hydroxyphenyl) prop-2-enoyl]-6-methyl-2*H*-pyran-2-one (Schiff base) and its copper complex (**287**) showed in vitro antifungal activity against a number of fungal species such as *A. flavus*, Curvularia *lunata*, and *Penicillium notatum*. These results indicated that the copper complex was having more inhibitory activity as compared with the free ligand [176].

287

5.6.5 DHA Thiosemicarbazone and its Ruthenium Complexes

DHA thiosemicarbazone (**288**, **289**) and its ruthenium complexes showed antifungal activity against the fungal species *Candida albicans* and *A. niger*. The most popular fungicide, namely clotrimazole (10 μg) (**290**), was used as a reference antibiotic. A greater antifungal activity has been observed by the above complexes against *A. niger* than against *C. albicans*. Even though all the metal complexes are active, they did not reach the efficiency of the fungicide clotrimazole [177]. These results indicate a comparison between ruthenium complexes and the free ligand.

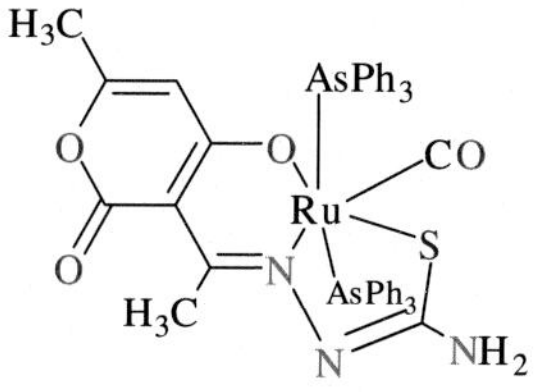

288

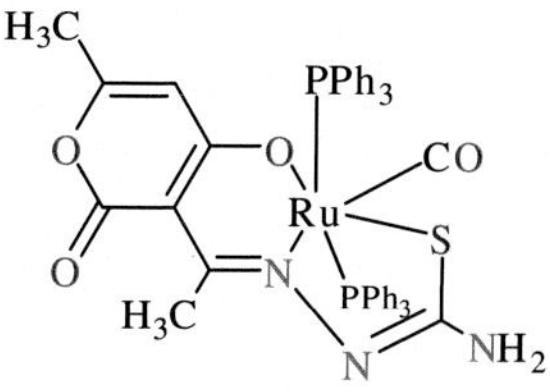

289

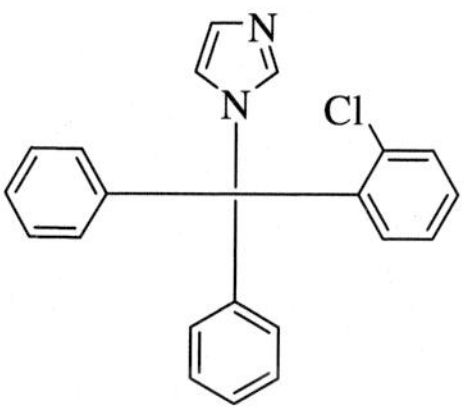

290

5.6.6 3-((*E*)-1-(2-((*E*)-4-Chloro/Fluorobenzylideneamino) Phenylimino)Ethyl)-4-Hydroxy-6-Methyl-2*H*-Pyran-2-Ones (Ligands) and Their Metal Complexes

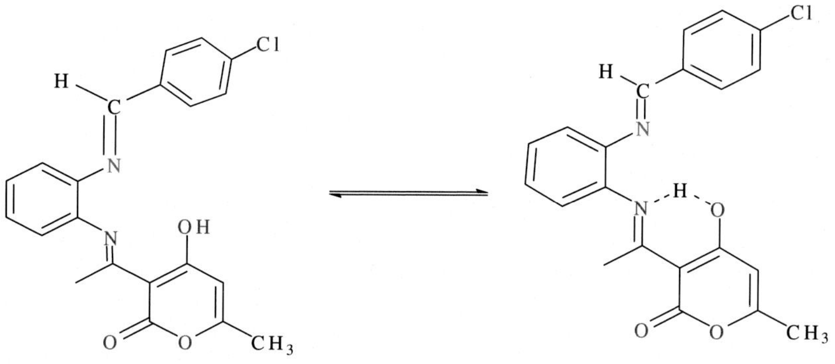

291

The antifungal activity of the compound chlorobenzylidene aminophenylimino derivative of DHA (Schiff base) (**291**) and its Cu, Ni, Co, Fe, and Mn complexes were examined against *Trichoderma* and *A. niger* using the mycelia dry weight method and ciprofloxin as a control or reference drug. The activity of these complexes is exactly similar to their stability constant order which is **Copper > Nickel > Cobalt > Iron > Manganese**. The above results of the ligand and its various metal complexes indicate that the **Cu** complex was nearly 4–5 times more biologically active against *A. niger*. The ligand activity was increased by chelating with Copper (II), Manganese (II), Cobalt (II), Nickel (II), and Ferrous (III) against *Trichoderma* [149]. The chloro substituent in the ligand is responsible for the high antifungal activity of the ligand and its metal complexes.

The antifungal activity of the compound fluorobenzylidene aminophenylimino derivative of DHA (Schiff base) (**292**) and its Cu, Ni, Co, Fe, and Mn metal complexes (**293**) were screened against the fungal species *Trichoderma* and *A. niger*. The inhibitory action of these metal complexes increases with the increase in the stability of the complex [195].

292   293

5.6.7 DHA and its DHT, Br DHA Analogs

DHA (**1**) and its two analogs, namely **DHT** (**294**) and **BrDHT** (**295**), are used to protect stored grains (soybeans). These are mainly used to prevent the production of aflatoxins (AFB$_1$, naturally occurring mycotoxins) and the impact of accumulation of AFB$_1$ (**286**) by strain *A. flavus* ATCC 26949 [196].

The application of the above DHA analogs on soybeans prior to storage is a potential means of preventing the growth and AFB$_1$ accumulation by *A. flavus* ATCC 26949.

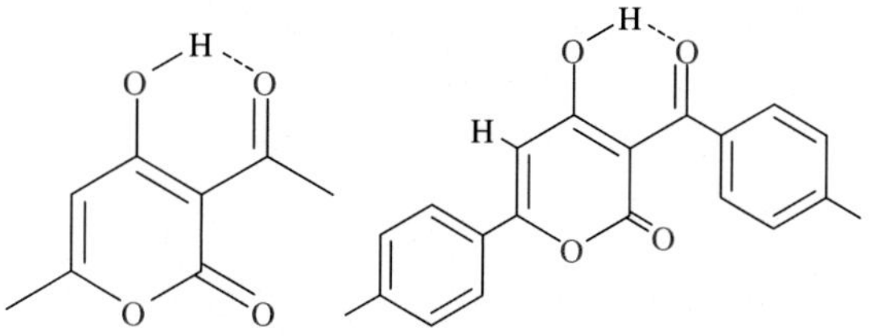

294

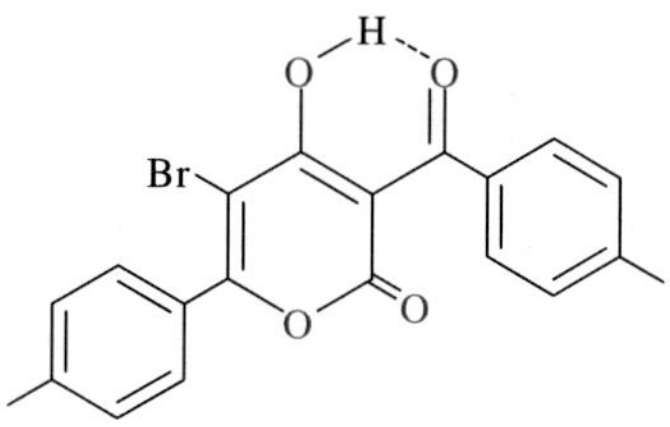

295

5.6.8   (*E*)-6-Methyl-3-(1-(Phenylimino)Ethyl)-2*H*-Pyran-2,4-Diol and its Metal Complexes

The antifungal activity of (*E*)-6-methyl-3-(1-(phenylimino)ethyl)-2*H*-pyran-2,4-diol (Schiff base) and its (**Ni** and **Co**) metal complexes (**296** and **297**) were assessed against four fungal strains, *A. niger*, *A. flavus*, *Fusarium moniliforme* [197] (Fig. 5.14), and *Penicillium chrysogenum* using penicillin and griseofulvin (**298**) as standard or reference drugs [154]. The structures of both Ni (II) and Co (II) metal complexes and various derivatives are given in Tables 5.1 and 5.2.

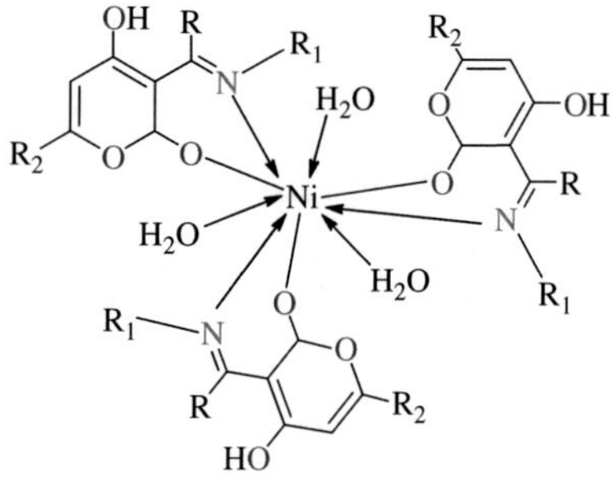

296

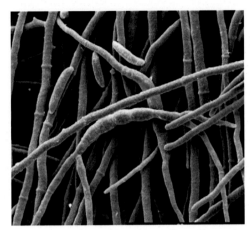

FIGURE 5.14   Scanning electron microscope images of *F. moniliforme* treated with 1% DMSO. *Kumar A, Lohan P, Aneja DK, Gupta GK, Kaushik D, Prakash O. Design, synthesis, computational and biological evaluation of some new hydrazino derivatives of DHA and pyranopyrazoles. Eur Jour Med Chem 2012;50:81–9.*

TABLE 5.1 Various Derivatives of Ni (II) Metal Complexes

S. No	R	R_1	R_2
a	CH_3	CH_3	![benzene-Cl]
b	CH_3	CH_3	![naphthalene]
c	CH_3	CH_3	![benzene]

TABLE 5.2 Various Derivatives of Co (II) Metal Complexes

S. No	R	R_1	R_2
a	CH_3	CH_3	![benzene-Cl]
b	CH_3	CH_3	![naphthalene]
c	CH_3	CH_3	![benzene]

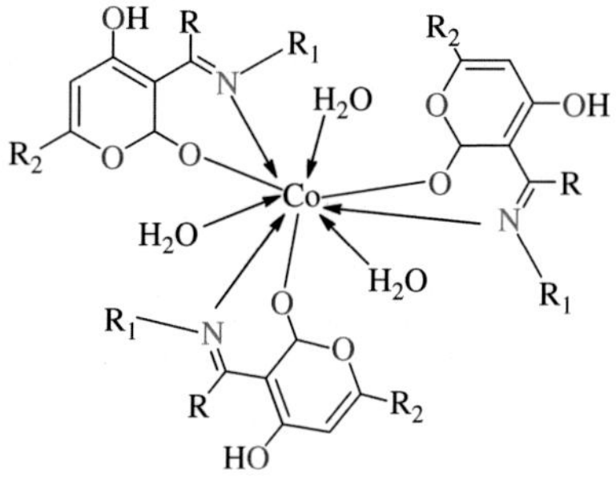

297

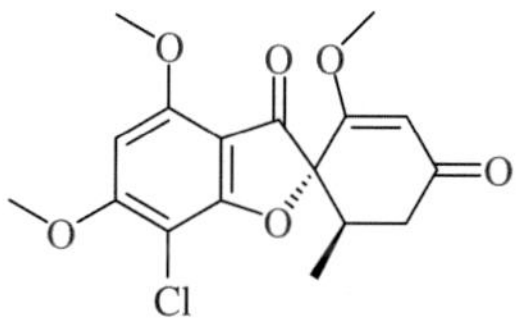

298

The complexes show higher antimicrobial activity than the ligands. This activity is due to the combined activity of the metal and ligand or faster diffusion of metal complexes through the cell membrane.

5.6.9 Chlorophenyl Substituted DHA

The antifungal activity of the compound (*E*)-3-(3-(2-chlorophenyl)acryloyl)-4-hydroxy-6-methyl-2*H*-pyran-2-one (**299**) against the species *P. chrysogenum*, *A. niger*, *A. flavus*, and *Fusarium moneliform* was assayed by using the plate method. Griseofulvin (**298**) was used as a reference drug and the above compound showed moderate activity against *A. niger* [198].

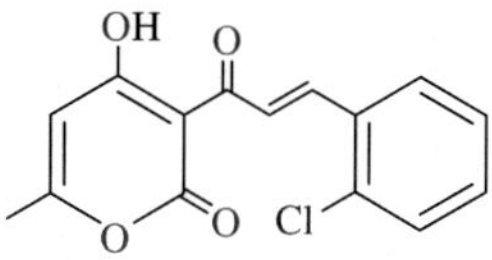

299

5.6.10 Azanediyl-bis-Ethanediyl-bis-Azanylylidene-Ethanylylidene-Hydroxyl Methyl Pyran-2-One Schiff Base and its Cu Metal Complex

Synthesized Cu (II) (**300**) complex was tested for antifungal activities by using Ketoconazole (antifungal agent) (**301**) as a standard drug. It was found that the

Cu (II) complex proved to be a potentially better antifungal agent against *A. flavus* than the Ketoconazole (standard) and this complex has better antimicrobial activity than the free ligand [199].

300

301

5.7 ANTIINFLAMMATORY ACTIVITY

The derivatives of DHA showed antiinflammatory activity against inflammation (the complex biological response of body tissues to destructive stimuli such as pathogens, damaged cells, or irritants). Inflammation is a protective response from immune cells, molecular mediators, and blood vessels. Inflammation is mandatory for eliminating the initial cause of cell injury, cleaning up necrotic cells, and initiating tissue repair. The usual signs of acute inflammation are pain, redness, heat, swelling, and loss of function. Inflammation is a general response and therefore it can be considered as a mechanism of innate immunity as compared to adaptive immunity which is peculiar for each pathogen.

5.7.1 Hydrazinyl DHA Derivatives and Pyranopyrazoles as Antiinflammatory Agents

The antiinflammatory activity of hydrazinyl DHA derivatives (**302** and **303**) and pyranopyrazoles (**304** and **305**) was screened via carrageenan-induced rat hind paw edema technique [200]. The compounds (**302, 303, 304**, and **305**) were bonded at the II pocket of the enzyme cyclooxygenase. These compounds have shown strong antiinflammatory activity that is almost the same as a standard drug diclofenac sodium [201] (**306**). The docking models of binding modes of the compounds to cyclooxygenase pocket are shown in Fig. 5.15.

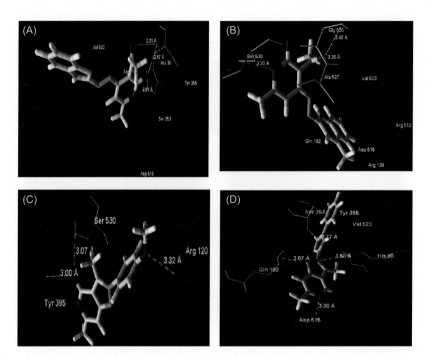

FIGURE 5.15 (A) Binding of the compound (**254**) into the cyclooxygenase II. (B) Binding of the compound (**255**) into the cyclooxygenase II. (C) Binding of the compound (**256**) into the cyclooxygenase II. (D) Binding of the compound (**257**) into the cyclooxygenase II. *Mahdieh M, Yazdani M, Mahdieh S. The high potential of Pelargonium roseum plant for phytoremediation of heavy metals. Environ Monit Assess 2013: doi:10.1007/s10661-013-3141-3.*

302

303

304

305

306

FIGURE 5.16   Manganese phytotoxicity on leaf. *Pal R, Kumar V, Gupta AK, Beniwal V. Synthesis, characterization and DNA photocleavage study of a novel dehydroacetic acid based hydrazone Schiff's base and its metal complexes. Med Chem Res 2014;23:3327–35.*

5.8   PHYTOTOXIC ACTIVITY

Phytotoxicity or phytotoxic activity is the measure at which a chemical or compound becomes harmful or lethal to plants [202]. All types of pesticides, especially herbicides, can be hazardous to plants because they are designed to kill or injure or suppress the growth of plants. Some types of insecticides and fungicides can also cause harm to plants (Fig. 5.16).

5.8.1   Enaminopyran-2,4-Dione as a Phytotoxic Agent

The phytotoxic activities of chiral (*E*)-enaminopyran-2,4-diones were assessed by their capability to inhibit the development of *Sorghum bicolour* (commonly called as *Sorghum*) and *Cucumis sativus* seedlings.

The most active chiral alkyl enamine compound, namely (*S*)-butylaminoethylidene-(methyl-pyran)-dione (**307**), was examined against two weeds, namely *Ipomoea grandifolia* and *Brachiaria decumbens*. This compound exhibited the maximum inhibitory action on the growth of aerial parts (71%) and roots (66%) of *B. decumbens* [203].

307

5.9 ANALGESIC ACTIVITY

Analgesics belong to a special group of drugs which are meant for giving relief from pain and inflammation without causing the loss of consciousness [204]. The different categories of analgesic drugs are:

- Nonsteroidal antiinflammatory drugs (NSAIDs), such as naproxen (with brand names as Aleve and Naprosyn), ibuprofen (with brand names as Advil and Motrin), or prescription Cox-2 inhibitors (with brand name as Celebrex).
- Narcotics and synthetic narcotic drugs, such as morphine and methadone, can also be used to relieve pain.

5.9.1 Hydrazinyl DHA Derivatives and Pyranopyrazoles as Analgesic Agents

The analgesic activity of hydrazino derivatives of DHA, namely 4-aryl/heteroaryl hydrazino-3-acetyl-6-methyl-2*H*-pyran-2-ones (**308–315**), was evaluated by following the tail immersion and acetic acid-induced writhing assay methods using diclofenac sodium as a standard or reference drug. These compounds show excellent analgesic activity compared with diclofenac sodium [201] (**306**).

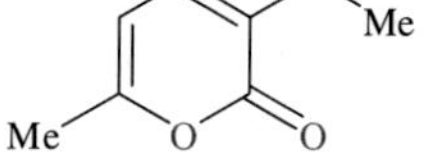

308

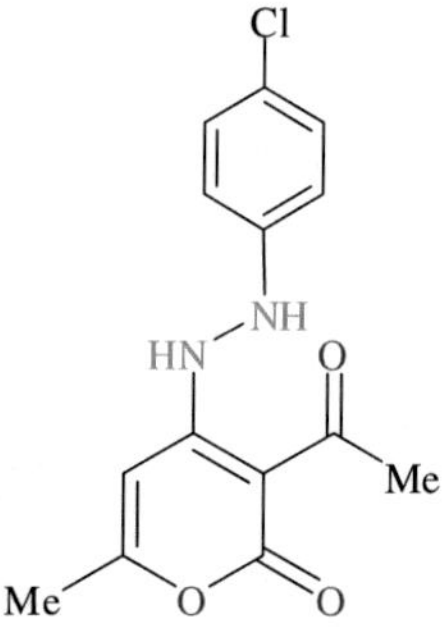

309

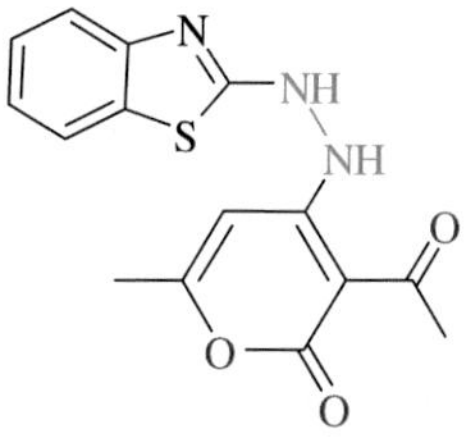

310

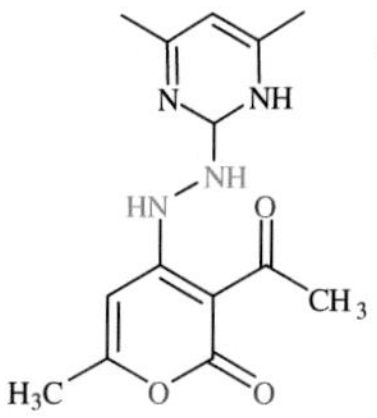

311

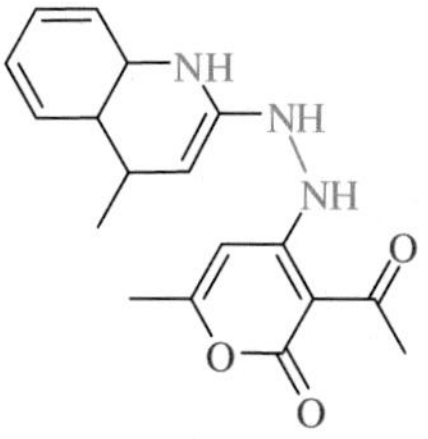

312

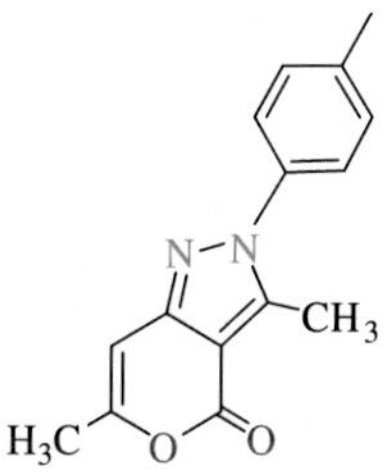

313

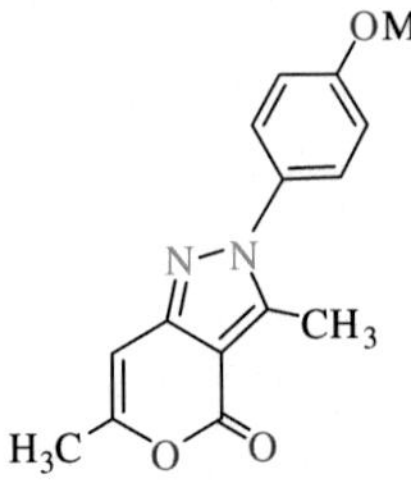

314 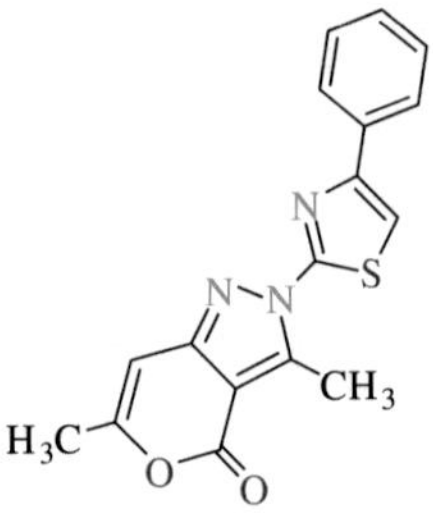   315

5.10 PHOTOCLEAVAGE ACTIVITY

5.10.1 Benzohydrazide Derivative of DHA and its Ni (II), Co (II), and Cu (II) Metal Complexes

The benzohydrazide derivative of DHA and its Ni (II), Co (II), and Cu (II) complexes (**316**) possess good DNA photocleavage activity [205].

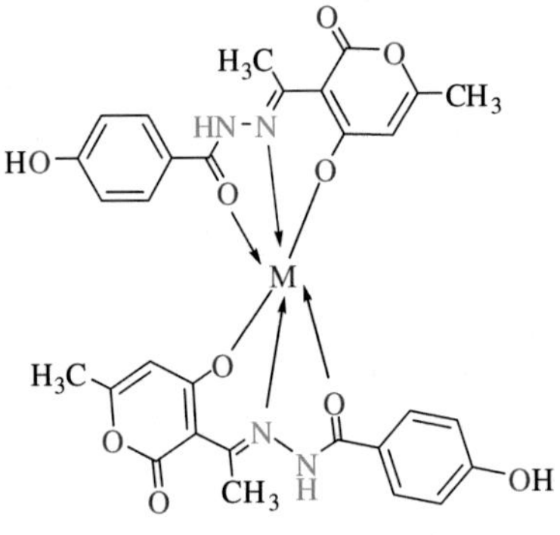

316

The Co (II) and Cu (II) complexes showed good levels of DNA photocleavage activity. The Ni (II) complex was found to be the least potent under identical experimental conditions. The nuclease activity of these complexes was found to be greater than the free ligand.

5.10.2 Cyanoacetohydrazide Derivative of DHA and its Metal Complexes

The cyano acetohydrazide derivative of DHA (ligand) (**317**) and its **Cu (II)**, **Co (II)**, **Ni (II)**, **Zn (II)**, and **Mn (II)** (**318–320**) metal complexes play a vital role in

TABLE 5.3 DNA Photocleavage

Line C	(DNA+DMSO)
Line 1	(DNA+ligand)
Line 2	(DNA+LCo)
Line 3	(DNA+LNi)
Line 4	(DNA+LCu)
Line 5	(DNA+LMn)
Line 6	(DNA+LZn)
Line 7	(DNA+L$_2$Co)
Line 8	(DNA+L$_2$Ni)
Line 9	(DNA+L$_2$Co)
Line 10	(DNA+L$_2$Mn)
Line 11	(DNA+L$_2$Zn)

DNA photocleavage reactions. In the case of the resulting compounds (Table 5.3, lines 1–11), there was a formation of smearing due to cleavage of DNA into small fragments [206].

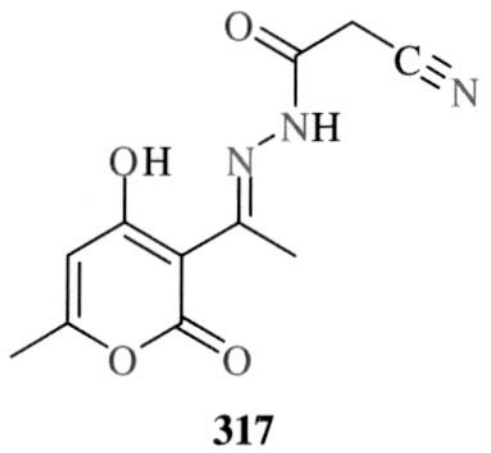

317

Therefore, the data assumed that the resulting complexes are excellent agents for DNA photocleavage. But in the case of metal complexes, the notable increase in cleaving ability was greater as compared to the nonbonded ligand because of their noncovalent interactions with genetic material, which again resulted in its cleavage (Fig. 5.17). Bands in the picture of DNA photocleavage are explained in Table 5.3.

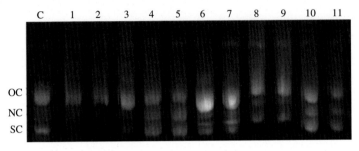

FIGURE 5.17   DNA photocleavage.

(LM) M = Cu (II),
NiMn (II), and Mn (II)

318

(LM) M = Zn (II) and Co (II)

319

(L₂M) M = Cu (II), Colt (II), Zn (II),and Ni (II)

320

The DNA photocleavage activity of both 1:1 and 1:2 complexes was similar without any difference. The L₂MCo was observed to be more active as compared to LMCo. The variations in the cleavage capacity of metal complexes were due to their unusual degree of binding abilities.

References

[1] Guo Y, Kannan K. A survey of phthalates and parabens in personal care products from the United States and its implications for human exposure. Environ Sci Technol 2013;47:14442–14449.

[2] Nair MSR, Cary ST. Metabolites of pyrenomycetes: XII. Polyketides from the hypocreales. Mycologia 1979;71:1089–96.

[3] Ohno H, Saheki T, Awaya J, Nakagawa A, Omura S. Isolation and characterization of elasnin, a new human granulocyte elastase inhibitor produced by a strain of Streptomyces. J Antibiot 1978;31:1116–23.

[4] Rivera C, Pinero E, Giral F. Dehydroacetic acid in anthers of Solandra nitida (Solanaceae). Experientia 1976;32:1490.

[5] Kheirandish R, Nourollahi-Fard SR, Yadegari Z. Prevalence and pathology of coccidiosis in goats in southeastern Iran. J Parasit Dis 2014;38:27–31.

[6] Geuther A, 2. Chem. (Jena), 2, 8 (1866); Chem. Zentr., 11, 801 (1866).

[7] Fiest F. Ueber Dehydracetsäure. Liebigs Ann 1890;257:253–97.

[8] (a) Collie On the constitution of dehydroacetic acid. J Chem Soc 1891;59:179–89.
    (b) Collie JN, Hilditch TP. An isomeric change of dehydroacetic acid. J Chem Soc 1907;91:787–9.

[9] Rassweiler CF, Adams R. The structure of dehydroacetic acid. J Am Chem Soc 1924;46:2758–64.

[10] Gupta DR, Gupta RS. J Indian Chem Soc 1965;42:421.

[11] Chalaca MZ, Figueroa-Villar JD. A theoretical and NMR study of the tautomerism of dehydroacetic acid. J Mol Struct 2000;554:225–31.

[12] Berson JA. On the structure of dehydroacetic acid. J Am Chem Soc 1952;74:5172–5.

[13] Gupta DR, Gupta RS. J Indian Chem Soc 1966;43:377–84.

[14] Yamada K. Infrared and ultraviolet spectra of α- and γ-pyrones. Bull Chem Soc Jpn 1962;35:1323–9.

[15] Talapatra SK, Basak A, Maiti BC, Talapatra B. 2,7-Dimethyl-4H-pyrano [3,2-C]-2H-pyran-4,5-dione, a novel product in the base-catalyzed self-condensation of ethyl acetoacetate-^{13}C NMR spectral studies. Indian J Chem 1980;19B:546–8.

[16] Reed RI, Takshishtov VV. Electron impact and molecular dissociation—XVI: β-keto esters. Tetrahedron 1967;23:2807–15.

[17] Duisberg C, Hess K, Leben AGS, Arbeiten S. Beric. der Deuts. Chemi Gesell 1930:145–57.

[18] Arndt F. Organic synthesis collective, vol. 20. New York: John Wiley & Sons, Inc; 1943, p. 26–7.

[19] Wang W. HuaxueShijie 1983;24 ChemAbstr 100 (1984) 34375a.

[20] Rauscher K, Wolfgang HO. Ardelt Ger (East) 1957;13:889. Chem Abstr 53 (1959) P 13179i.

[21] Balenovic K. Rec Trew Chim 1948;67:282. Chem Abstr 42 (1948) 8161i.

[22] Miyaki K, Yamagichi S. J Pharm Soc Jpn 1953;73:982. Chem Abstr 48 (1954) 4527d.

[23] Montagna AE, Lashley ER. V.S. 2912441; 1959; Chem Abstr 54 (1960) P 3457f.

[24] Isoshima T. Nippon Kagaku Zashi 1956;79:840–3. Chem Abstr 54 (1960) 4552.

[25] Marcus E, Chan KJ. Br Pat 1971;224:642. Chem Abstr 75 (1971) 5101n.

[26] Steele AB, Borse AB, Dull MF. J Org Chem 1969;14:460.

[27] Nakamura S, Ishidoya S. Nippon Synth Chem Jpn 1955:1884. Chem. Abstr. 51 (1956) 4442 d.

[28] Stephen JF, Marcus E. Reactions of dehydroacetic acid and related pyrones with secondary amines. J Org Chem 1969;34:2527–34.

[29] Wiley RH, Jarboe CH, Ellert HG. 2-Pyrones. XV. Substituted 3-cinnamoyl-4-hydroxy-6-methyl-2-pyrones from dehydroacetic acid. J Am Chem Soc 1955;77:5102–5.

[30] Mahesh VK, Gupta RS. Condensation products of dehydracetic acid with aromatic aldehydes and pyrazolines derived from them. Indian J Chem 1974;12:956.

[31] Richedi Y, Hamdi M, Speziale Y. Synthesis of 4-hydroxy 6-methyl 3-β-arylpropionyl 2-pyrones by selective catalytic hydrogenation of 3-cinnamoyl 4-hydroxy 6-methyl 2-pyrones. Synth Commun 1989;19:3437–42.

[32] Richedi Y, Hamdi M, Sakellariou R, Speziale Y. Reaction of 4-hydroxy-6-methyl-3-β-arylpropionyl-2-pyrones with phenylhydrazine-synthesis of a new pyrazole series. Synth Commun 1991;21:1189–99.

[33] Ichihara A, Miki M, Tazaki H, Sakamura S. Synthesis of (±)-solanapyrone A. Tetrahedron Lett 1987;28:1175–8.

[34] Lowe W. 4-Hydroxy-5-oximino-7-methyl-5H-pyrano[2,3-b] pyridin-8-oxid. Arch Pharm (Weinheim Ger) 1978;311:414–20.

[35] Harris TM, Harris CM, Brush CK. Bromination of dehydroacetic acid. J Org Chem 1970;35:1329–33.

[36] Hirsh B, Hoefgan N, Hashi S, Ali K. German patent, 238048; 1986; Chem Abstr 107 (1987) 115489 m.

[37] Hassan MA, Kady MEI, Abid Mohey AA. Indian J Chem 1982;21B:372.

[38] Akhrem AA, Moisenkov AM, Lakhwich FA, Kadentsev VI. Izv Akad Nauk SSSR Ser Khim 1970;5:1206. Chem Abstr 73 (1970) 66490.

[39] March P, Marquet J, Manas MM, Plexiats R, Ripoll I. A. Trins, metal complexes in organic synthesis VII. Alkylation of o-dicarbonyl compounds, phenols and pyrones with allylic and benzylic alcohols under cobalt (ii) chloride catalysis. An Quim 1983;79C:15.

[40] Marquet J, Manas MM. Alkylation of active hydrogen compounds with allylic and benzylic alcohols under $CoCl_2$ catalysis. A useful synthesis of Grifolin. Chem Lett 1981;10:173–6.

[41] Bacardit R, Manas MM, Plexiats R. Functionalization at C-5 and at the C-6 methyl group of 4-methoxy-6-methyl-2-pyrone. J Heterocycl Chem 1982;19:157–60.

[42] Bacardit R, Cerevello J, de March P, Marquet J, Manas MM, Roca JL. Chemistry of pyrones related to dehydroacetic acid. Functionalization at C-5 and at the methyl group at C-6. An attempted synthesis of a thromboxane B_2 analogue. J Heterocycl Chem 1989;26:1205–12.

[43] Harris TM, Harris CM, Wachter MP. Condensations of dehydroacetic acid at the 6-methyl position. Tetrahedron 1968;24:6897–906.

[44] Bacardit R, Manas MM. A very simple synthesis of natural saturated Δ-substituted Δ-lactones. The pheromone of vespa orientalis. Chem Lett 1982;11:5–6.

[45] Bacardit R, Manas MM. Synthesis of δ-lactonic pheromones of Xylocopa hirsutissima and Vespa orientalis and an allomone of some ants of genus Camponotus. J Chem Ecol 1983;9:703–14.

[46] Deisig N, Dupuy F, Anton S, Renou M. Responses to pheromones in a complex odor world: sensory processing and behavior. Insects 2014;5:399–422.

[47] Nedjar-Kolli B, Hamdi M, Piere J, Harault V. Structure et synthèses nouvelles dans la série des amino-4 dihydro-5,6 méthyl-6 pyrones-2. J Heterocycl Chem 1981;18:543–7.

[48] Ayer WA, Villar JDF. Metabolites of Lachnellula fuscosanguinea (Rehm). Part 1. The isolation, structure determination, and synthesis of lachnelluloic acid. Can J Chem 1985;63:1161–5.

[49] Collie JN. The lactone of triacetic acid. J Chem Soc 1891;59:607–17.

[50] Rohn. Naas Japan patent 58, 216183; 1983; Chem Abstr 100 (1984) 156497.

[51] Collie N, Myers WS. The formation of orcinol and other condensation products from dehydracetic acid. J Chem Soc 1893;63:122–8.

[52] Rama Rao N, Rao DS, Ganorkar MC. Indian J Chem 1982;21A:839.

[53] Rama Rao N, Rao PV, Reddy GV, Ganorkar MC. Metal complexes of physiologically active O:N:S tridentate Schiff base. Indian J Chem 1987;26A:887–90.

[54] Rama Rao N, Ganorkar MC. Indian J Chem 1988;27A:52.

[55] Prasad KM, Rama Rao N, Ganorkar MC. Synthesis, spectra and biological activity of some new telluronium heterocyclic β-diketonates. Organomet Chem 1989;376:53–9.

[56] Surya Rao D. Studies on some transition metal complexes synthesized from physiologically active ligands. Ph.D. Thesis, Osmania University; 1981.

[57] Vijay Krishna Kumar K, Reddy MS, Kloepper JW, Lawrence KS, Zhou XG, Groth DE, et al. Commercial potential of microbial inoculants for sheath blight management and yield enhancement of rice Bacteria in agrobiology: crop ecosystems. New York: Springer; 2011.237.63

[58] Valent B. Underground life for rice foe. Nature 2004;431:516–7.

[59] Meiki A, Matsuda M, Watanabe H, Sawatu M, Iwataki I. Chem Abstr 1976;85:15368.

[60] Poulton GA, Cyr TD. Pyrones. IX. Synthetic approaches to the fungal metabolite phacidin and its derivatives. Can J Chem 1982;60:2821–9.

[61] Tanabe Y, Miyakada M, Ohno M, Hirosuke YN. Chem Lett 1982:1543.

[62] Vleggar P. Biosynthetic studies on some polyene mycotoxins. Pure Appl Chem 1986;58:239–56.

[63] Jones G. Sandoz patent Germany 3, 820, 538; 1989; Chem Abstr 110 (1989) 212625x.

[64] Kingigoshi U. Chemotherapy 1960;8:84. Chem Abstr 54 (1960) 22844a.

[65] Masaji O, Musato M, Matsunga K, Tadashi S. Japan patent 01, 261, 300; 1989; Chem Abstr 113 (1990) 126597v.

[66] Omura S, Ohno H, Saheki T, Masazu Y, Nakagawa A. Elasnin, a new human granulocyte elastase inhibitor produced by strain of Streptomyces. Bio-Chem Bio-Phys Res Commun 1978;83:704–9.

[67] Mors WB, Magalhaes MT, Gottlieb VR. Naturally occurring aromatic derivatives of monocyclic α-pyrones. Fortschr Chem Org Naturst 1962;20:131–64.

[68] Rehse K, Ruther D. Einfluß der S-Oxidation auf anticoagulante Wirkungen bei 4-Hydroxycumarinen, 4-Hydroxy-2-pyronen und 1,3-Indandionen. Arch Pharm (Weinheim Ger) 1984;317:262–7.

[69] Sekiguchi T, Nakasawa K. Japan kokai 74, 144, 449; 1979; Chem Abstr 92 (1990) 182094p.

[70] Grinblat MP, Rozova NJ, Kate IA, Spiridinova SA. Kauch Rezina 1975;11:9. Chem Abstr 84 (1976) 91332.

[71] Rapp GW. U.S. patent 2, 791, 590; 1957; Chem Abstr 51 (1957) 15594h.

[72] Agarwal R, Rani C, Sharma C, Aneja KR. Methodology synthesis and biological evaluation of new 1-(4,6-dimethylpyrimidin-2-yl)-1'-aryl/heteroaryl-3,3'-dimethyl-(4,5'-bipyrazol)-5-ols as antimicrobial agents. Rep Org Chem 2013;3:13–22.

[73] Cantos A, March P, Moreno-Manas M, Pla A, Sanchez-Fernando F, Virgili A. Synthesis of pyrano[4,3-c]pyrazol-4(1H)-ones and -4(2H)-ones from dehydroacetic acid. Homo- and heteronuclear selective NOE measurements for unambiguous structure assignment. Bull Chem Soc Jpn 1987;60:4425–31.

[74] Gupta DR, Ojha AC. Condensation products of some substituted aromatic amino compounds with dehydroacetic acid. J Indian Chem Soc 1971;48:291.

[75] Gupta DR, Ojha AC. J Indian Chem Soc 1970;47:1207.

[76] Qayyum MA, Hanumanthu P, Ratnam CV. Indian J Chem 1982;21B:883–5.

[77] Strakov A, Yasuhu M, Egle A, Mols A. Latv PSR Zinat Akad Vestis Khim Ser 1970;5:615.

[78] Liu S, Rettig SJ, Orvig C. Synthesis and characterization of a pentadentate Schiff base N_3O_2 ligand and its neutral technetium (V) complex. X-ray structure of {N,N'-3-azapentane-1,5-diylbis(3-(1-iminoethyl)-6-methyl-2H-pyran-2,4(3H)-dionato) (3-) -O,O',N,N',N"} oxo-technetium (V). Inorg Chem 1991;30:4915–9.

[79] Elguero J, Martinez A, Singh SP, Grover M, Tarar LS. *A ^1H and ^{13}C NMR study of the structure and tautomerism of 4-pyrazolylpyrazolinones.* J Heterocyl Chem 1990;27:865–70.

[80] Gelin S, Chatrengel B, Naidi AI. Synthesis of 4-(acylacetyl)-1-phenyl-2-pyrazolin-5-ones from 3-acyl-2H-pyran-2,4(3H)-diones. Their synthetic applications to functionalized 4-oxopyrano[2,3-c]pyrazole derivatives. J Org Chem 1983;48:4078–82.

[81] Bennamane N, Nedjar-Kolli B, Geronikaki AA, Eleftheriou P, Kaoua R, Boubekeur K, et al. N-Substituted [phenyl-pyrazolo]-oxazin-2-thiones as COX-LOX inhibitors: influence of the replacement of the oxo -group with thioxo- group on the COX inhibition activity of N-substituted pyrazolo-oxazin-2-ones. Arkivoc 2011:69–82.

[82] Haitinger L. Ueber die Dehydracetsäure. Chem Ber 1885;18:452–3.

[83] Wang CS, Easterly JP, Skelly NF. Reaction of Dehydroacetic acid with ammonia. Tetrahedron Lett 1971;27:2581–9.

[84] Sammes MP, Yip KL. J Chem Soc Perkin Trans 1 1978:1373.

[85] Cook D. The preparation, properties, and structure of 2,6-bis-(alkylamino)-2,5-heptadien-4-ones. Can J Chem 1963;41:1435–40.

[86] Garrat S. The mechanism of the reaction between dehydroacetic acid and alkylamines. J Org Chem 1963;28:1886–8.

[87] Chang L, Sha J, Fu H. Chinese patent 1, 044, 462; 1990; Chem Abstr 116 (1992) 20949h.

[88] Afridi AS, Katrizky AR, Ramsdan CA. Preparation of NN'-linked bi(heteroaryls) from dehydroacetic acid and 2,6-dimethyl-4-pyrone. J Chem Soc Perkin Trans 1 1977:1428–36.

[89] Sammes MP, Wah HK, Katrizky AR. J Chem Soc Perkin Trans 1 1977:327–32.

[90] Castellanos ML, et al. J Chem Soc Perkin Trans 1 1985:1209–15.

[91] Akheram AA, Moisenkov AM, Krivoruchko VA. Approach to synthesis of 8-azasteroids Communication 3. Condensation of 3,4-dihydroisoquinolines with derivatives of dehydroacetic acid. Ser Khim 1973;22:1258–62.

[92] Hubert Habart M, Pene C, Royer R. Research for antitumor agents. 7. Possibility of producing pyrimidines by nucleophilic attack on 4-hydroxy alpha-pyrones. Chim Ther 1973;8:194–8.

[93] Echert K, Paul EF. Ger. Offen. 2, 220, 181; 1973; Chem Abstr 80 (1974) 27281f.

[94] Takahiro K, Hideo O, Toshihisa W, Tadashi N. European patent 347,866; 1989; Chem Abstr 113 (1990) 23941y.

[95] Batelaan JG. A convenient synthesis of triacetic acid methyl ester. Synth Commun 1976;6:81–3.

[96] Cerevello J, Marquet J, Manas MM. Chem Commun 1987:644.

[97] Cerevello J, Marquet J, Manas MM. Copper complex protection in the regioselective alkylation of methyl 3,5-dioxohexanoate. Preparation of 3-alkyl derivatives of 4-hydroxy-6-methyl-2-pyrone. Tetrahedron Lett 1987;28:3715–6.

[98] Marquet J, Manas MM, Prat M. Double bond formation by one pot palladium induced reactions between aldehydes, allylic alcohols and triphenylphosphine. Tetrahedron Lett 1989:3109–12.

[99] Cerevello J, Marquet J, Manas MM. Copper and cobalt mediated regioselective alkylation of polyketide models: methyl 3,5-dioxohexanoate and triacetic acid lactone. Tetrahedron Lett 1990;46:2035–46.

[100] Bethell JP, Maitland P. Organic reactions in aqueous solution at room temperature. Part III. The influence of pH on the self-condensation of diacetyl-acetone: constitution of Collie's naphthalene derivative. J Chem Soc 1962:3751–8.

[101] Abbasi EM, Djerrari B, Essassi EM, Fifani J. L'acide dehydracetique, precurseur de synthese de benzodiazepines. Tetrahedron Lett 1989;30:7069–70.

[102] Redha Al-Bayati IH, Sajida M, Thamir A. Mustansiriyah, synthesis of novel compounds derived from dehydroacetic acid. Al-Mustansiriyah J Sci 2013;24:41–56.

[103] Aït-Baziz N, Rachedi Y, Silva AMS. Reactivity of some structural analogs of dehydroacetic acid with o-phenylenediamine. Arkivoc 2010:86–97.

[104] Prasad YR, Rajasekhar KK, Bhaskar Rao B, Shankarananth V, Murthy K. Solvent free microwave assisted condensation of dehydroacetic acid with aromatic and heteroaromatic aldehydes. Int J Innov Pharm Res 2010;1:44–7.

[105] Wanga H, Zou Y, Zhao X, Shi D. A novel and convenient synthesis of 4-hydroxy-6-methyl-3-(1-(phenylimino)ethyl)-2H-pyran-2-one derivatives under ultrasound irradiation. Ultrason Sonochem 2011;18:1048–51.

[106] Prakash R, Kumar A, Singh SP, Aggarwal R, Prakash O. Dehydroacetic acid and its derivatives in organic synthesis: synthesis of some new 2-substituted-4- (5-bromo-4-hydroxy-6-methyl- 2H-pyran-2-one-3-yl)thiazoles. Indian J Chem 2007;46B:1713–5.

[107] Agarwal R, Rani C, Sharma C, Aneja KR. Synthesis and biological evaluation of new 1-(4, 6-dimethylpyrimidin-2-yl)-1'-aryl/heteroaryl-3,3'-dimethyl-(4,5'-bipyrazol)-5-ols as antimicrobial agents. Rep Org Chem 2013;3:13–22.

[108] Kaoua R, Bennamane N, Bakhta S, Benadji S, Rabia C, Nedjar-Kolli B. Synthesis of substituted 1,4-diazepines and 1,5-benzodiazepines using an efficient heteropolyacid-catalyzed procedure. Molecules 2011;16:92–9.

[109] Mokhtar F, Pascal H, Mohamed A. Synthesis of 4-pyrano-1,5-benzodiazepines catalysed by bismuth (III) derivatives. Res J Pharm Biol Chem Sci 2012;3:10–15.

[110] Prakash O, Kumar A, Sadana A, Prakash R, Singh SP, Claramunt RM, et al. Study of the reaction of chalcone analogs of dehydroacetic acid and o-aminothiophenol: synthesis and structure of 1,5-benzothiazepines and 1,4-benzothiazines. Tetrahedron 2005;61:6642–51.

[111] Aït-Baziz N, Rachedi Y, Chemat F, Hamdi M. Solvent free microwave-assisted knoevenagel condensation of dehydroacetic acid with benzaldehyde derivatives. Asian J Chem 2008;20:2610–22.

[112] Alvim HGO, da Silva Júnior EN, Neto BAD. What do we know about multicomponent reactions? Mechanisms and trends for the Biginelli, Hantzsch, Mannich, Passerini and Ugi MCRs. RSC Adv 2014;4:54282–54299.

[113] Strecker A. Ueber die künstliche Bildung der Milchsäure und einen neuen, dem Glycocoll homologen Körper. Justus Liebigs Ann Chem 1850;75:27–45.

[114] Hantzsch A. Neue bildungsweise von pyrrolderivaten. Ber Dtsch Chem Ges 1890;23:1474–6.

[115] Bade T, Rao VR. A facile one-pot synthesis of 3-(1-benzyl-2-phenyl-1H-imidazol-4-yl)-4-hydroxy-6-methyl-2H-pyran-2-one derivatives via multi-component approach. Org Commun 2014;7:53–9.

[116] Santhosh P, Rao VR. A facile one-pot synthesis of 1,3,4-thiadiazine-5-yl-pyran-2-one derivatives via a multicomponent reaction. J Sulfur Chem 2011;32:327–33.

[117] Santhosh P, Rao VR. New, convenient, one-pot method for the synthesis of thiazolyl-pyrazolones from dehydroacetic acid derivative via a multicomponent approach. Synth Commun 2012;42:3395–402.

[118] Santhosh P, Rao VR. Synthesis of 2,4,6-tri-substituted pyridine derivatives in aqueous medium via hantzsch multi-component reaction catalyzed by cerium (IV) ammonium nitrate. J Heterocycl Chem 2013;50:859–62.

[119]   Karkas MD, Akermark B. Water oxidation using earth-abundant transition metal catalysts: opportunities and challenges. Dalton Trans 2016;45:14421–14461.

[120]   Itodo AU, Usman A, Sulaiman SB, Itodo HU. Color matching estimation of iron concentrations in branded iron supplements marketed in Nigeria. Adv Anal Chem 2012;2:16–23.

[121]   Lia Y, Wua Y, Zhaoa J, Pin Y. DNA-binding and cleavage studies of novel binuclear copper(II) complex with 1,1'-dimethyl-2,2'-biimidazole ligand. J Inorg Biochem 2007;101:283–90.

[122]   Wang XL, Chao H, Hong L, Hong XL, Nian JL, Li XY. Synthesis, crystal structure and DNA cleavage activities of copper(II) complexes with asymmetric tridentate ligands. J Inorg Biochem 2004;98:423–9.

[123]   Cowan JA. Chemical nucleases. Curr Opin Chem Biol 2001;5:634–42.

[124]   Lee KB, Wang D, Lippard SJ, Sharp PA. Transcription-coupled and DNA damage-dependent ubiquitination of RNA polymerase II in vitro. Proc Natl Acad Sci USA 2002;99:4239–44.

[125]   Solomons NW. Biochemical, metabolic, and clinical role of copper in human nutrition. J Am Coll Nutr 1985;4:83–105.

[126]   Williams DR. The metals of life. London: Van Nostrand Reinhold; 1971.

[127]   Bugarcic ZD, Bogojeski J, Petrovic B, Hochreuther S, van Eldik R. Mechanistic studies on the reactions of platinum (II) complexes with nitrogenand sulfur-donor biomolecules. Dalton Trans 2012;41:12329.

[128]   Sorenson JRJ. Copper chelates as possible active forms of the antiarthritic agents. J Med Chem 1976;19:135–48.

[129]   Rademaker-Lakhai JM, Bongard DV, Pluim D, Beijnen JH, Schellens JHM. A phase I and pharmacological study with imidazolium-trans-DMSO-imidazole-tetrachlororuthenate, a novel ruthenium anticancer agent. Clin Cancer Res 2004;10:3717–27.

[130]   Hartinger CG, Jakupec MA, Zorbas-Seifried S, Groessl M, Egger A, Berger W, et al. KP1019, a new redox-active anticancer agent–preclinical development and results of a clinical phase I study in tumor patients. Chem Biodiversity 2008;5:2140–55.

[131]   Messori L, Abbate F, Marcon G, Orioli P, Fontani M, Mini E, et al. Gold (III) complexes as potential antitumor agents: solution chemistry and cytotoxic properties of some selected gold (III) compounds. J Med Chem 2000;43:3541–8.

[132]   Decker A, Chow MS, Kemsley JN, Lehnert N, Solomon EI. Direct hydrogen-atom abstraction by activated bleomycin: an experimental and computational study. J Am Chem Soc 2006;128:4719–33.

[133]   Hashimoto SE, Wang BX, Hecht SM. Kinetics of DNA cleavage by Fe (II) bleomycins. J Am Chem Soc 2001;123:7437–8.

[134]   Hsieh W, Zaleski CM, Pecoraro VL, Fanwick PE, Liu S. Mn(II) complexes of monoanionic bidentate chelators: X-ray crystal structures of Mn(dha)$_2$(CH$_3$OH)$_2$ (Hdha=Dehydroacetic acid) and [Mn(ema)$_2$(H$_2$O)]$_2$ 2H$_2$O (Hema=2-ethyl-3-hydroxy-4-pyrone). Inorg Chim Acta 2006;359:228–36.

[135]   Chatt J, Leigh GJ, Mingos DMP, Paske RJ. Complexes of osmium, ruthenium, rhenium, and iridium halides with some tertiary monophosphines and monoarsines. J Chem Soc 1968:2636–41.

[136]   Poddar RK, Khullar IP, Agarwala U. Some ruthenium (III) complexes with triphenylarsine. J Inorg Nucl Chem Lett 1974;10:221.

[137]   Natarajan K, Poddar RK, Agarwala U. Mixed complexes of ruthenium(III) and ruthenium(II) with triphenylphosphine or triphenylarsine and other ligands. J Inorg Nucl Chem 1977;39:431–5.

[138]   Ahmed N, Lewison JJ, Robinson SD, Uttley MF. Complexes of ruthenium, osmium, rhodium, and iridium containing hydride carbonyl, or nitrosyl ligands. Inorg Synth 1974;15:45–64.

[139] Delgado RAS, Lee WY, Choi SR, Cho Y, Jun MJ. The chemistry and catalytic properties of ruthenium and osmium complexes. Part 5. Synthesis of new compounds containing arsine ligands and catalytic activity in the homogeneous hydrogenation of aldehydes. Trans Met Chem 1991;16:241–4.

[140] Chitrapriya N, Mahalingam V, Zeller M, Jayabalan R, Swaminathan K, Natarajan K. Synthesis, crystal structure and biological activities of dehydroacetic acid complexes of Ru (II) and Ru (III) containing PPh$_3$/AsPh$_3$. Polyhedron 2008;27:939–46.

[141] Krepsky N, Ferreira RBR, Nunes APF, Lins UGC, Filho FCS, de Mattos-Guaraldi AL, et al. Cell surface hydrophobicity and slime production of staphylococcus epidermidis Brazilian isolates. Curr Microbiol 2003;46:280–6.

[142] Di Ciccio P, Vergara A, Festino AR Paludi D, Zanardi E, Ghidini S, et al. Biofilm formation by Staphylococcus aureus on food contact surfaces: relationship with temperature and cell surface hydrophobicity. Food Control 2014;50:930–6.

[143] Deldar AA, Yakhchali B. The influence of riboflavin and nicotinic acid on Shigella sonnei colony conversion. Iran J Microbiol 2011;3:13–20.

[144] Zdybicka-Barabas A, Mak P, Klys A, Skrzypiec K, Mendyk E, Fiołka MJ, et al. Synergistic action of Galleria mellonella anionic peptide 2 and lysozyme against gram-negative bacteria. Biochim Biophys Acta 2012;1818:2623–35.

[145] Fouad DM, Bayoumi A, El-Gahami MA, Ibrahim SA, Hammam AM. Synthesis and thermal studies of mixed ligand complexes of Cu (II), Co (II), Ni (II) and Cd (II) with mercaptotriazoles and dehydroacetic acid. Nat Sci 2010;2:817–27.

[146] Cansiz A, Koparir M, Demitdag A. Synthesis of some new 4,5-substituted-4H-1,2,4-triazole-3-thiol derivatives. Molecules 2004;9:204–12.

[147] Ragenovic KC, Dimova V, Kakurinov V, Molnar DG, Buzarovska A. Synthesis, antibacterial and antifungal activity of 4-substituted-5-aryl-1,2,4-triazoles. Molecules 2001;6:815–24.

[148] Reid JR, Heindel D. Improved syntheses of 5-substituted-4-amino-3-mercapto-(4H)-1,2,4-triazoles. J Heterocycl Chem 1976;13:925–6.

[149] Jadhav SM, Munde AS, Shankarwar SG, Patharkar VR, Shelke VA, Chondhekar TK. Synthesis, potentiometric, spectral characterization and microbial studies of transition metal complexes with tridentate ligand. J Korean Chem Soc 2010;54:515–22.

[150] Munde AS, Jagdale AN, Jadhav SM, Chondhekar TK. Synthesis, characterization and thermal study of some transition metal complexes of an asymmetrical tetradentate Schiff base ligand. J Serb Chem Soc 2010;75:349–59.

[151] Mane PS, Shirodkar SG, Arbad BR, Chondhekar TK. Synthesis and characterization of manganese(II), cobalt(II), nickei(II), and copper(II) complexes of Schiff base derivatives of dehydroacetic acid. Indian J Chem 2001;40:648–51.

[152] Al-Jibouri MN. Synthesis and structural studies of Oxo-Vanadium (IV), Chromium(III), Manganese(II), Iron(II), Cobalt(II), Nickel(II), and Copper(II) complexes with a new tetra dentate Schiff base having O: N: N: O donor system. J Al-Nahrain Univ 2008;11:10–15.

[153] Deshmukh PS, Yaul AR, Bhojane JN, Aswar AS. Synthesis, characterization and thermogravimetric studies of some metal complexes with N$_2$O$_2$ Schiff base ligand. World Appl Sci J 2010;9:301–1305.

[154] Habib SI, Kumar P, Kulkarni A. Synthesis and characterization of complexes of Schiff bases of transition metals. Int J Adv Pharm Biol Chem 2012;1:234–7.

[155] Maiti BC, Maitra SK. Reaction of dehydroacetic acid with aliphatic, aromatic and heterocyclic amines. Indian J Chem 1998;37:710–2.

[156] Cindric M, Vrdoljak V, Novak TK, Curic M, Brbot-Saranovic A, Kamenar B. Synthesis and characterization of two dehydroacetic acid derivatives and molybdenum(V) complexes: an NMR and crystallographic study. J Mol Struct 2004;701:111–8.

[157]   Iqbal J, Wattoo FH, Tirmizi SA, Wattoo MHS. Dehydroacetic acid oxime as a new ligand for spectrophotometeric determination of cobalt. J Chem Soc Pak 2007;29:136–9.

[158]   Kubaisi AL, Ismail K. Nickel (II) and palladium (II) chelates of dehydroacetic acid Schiff bases derived from thiosemicarbazide and hydrazinecarbodithioate. Can J Chem 1994;72:1785–8.

[159]   Benassi CA, Bettero A, Manzini P, Semenzato A, Traldi P. Interaction between dehydroacetic acid sodium salt and formaldehyde: structural identification of the product. J Soc Cosmet Chem 1988;39:85–92.

[160]   Al Alousi AS, Shehata MR, Shoukry MM, Hassan SA. Coordination properties of Dehydroacetic acid–binary and ternary complexes. J Coord Chem 2008;61:1906–16.

[161]   Rao PV, Narasaiah AV. Synthesis, characterization and biological studies of Oxovanadium(IV), Manganese(II), Iron(II), Cobalt(II), Nickle(II) and Copper(II) Complexes derived from a Quadridentate ligand. Indian J Chem 2003;42:1896–9.

[162]   Raman N, Johnson SR, Sukthicevel A. Transition metal complexes with Schiff-base ligands: 4-aminoantipyrine based derivatives-a review. J Coord Chem 2009;62:691–709.

[163]   Karlsson Hedestam GB, Fouchier RAM, Phogat S, Burton DR, Sodroski J, Wyatt RT. The challenges of eliciting neutralizing antibodies to HIV-1 and to influenza virus. Nat Rev 2008;6:143–55.

[164]   Pommier Y, Johnson AA, Marchand C. Integrase inhibitors to treat HIV/AIDS. Nat Rev 2005;4:236–48.

[165]   Barré-Sinoussi F, Ross AL, Delfraissy J-F. Past, present and future: 30 years of HIV research. Nat Rev 2013;11:877–83.

[166]   Tambov KV, Voevodina IV, Manaev AV, Ivanenkov YA, Neamati N, Travena VF. Structures and biological activity of cinnamoyl derivatives of coumarins and dehydroacetic acid and their boron difluoride complexes. Russ Chem Bull 2012;61:78–90.

[167]   Ku J-L, Park S-C, Kim K-H, Jeon Y-K, Kim S-H, Shin Y-K, et al. Establishment and characterization of seven human breast cancer cell lines including two triple-negative cell lines. Int J Oncol 2013;43:2073–81.

[168]   Kashar TI, El-Sehli AH. Synthesis, characterization, antimicrobial and anticancer activity of Zn(II), Pd(II) and Ru(III) complexes of dehydroacetic acid hydrazine. J Chem Pharm Res 2013;5:474–83.

[169]   Holmes JA, Chung RT. HCV compartmentalization in HCC: driver, passenger or both? Nat Rev 2016;13:254–6.

[170]   Noureini SK, Wink M. Transcriptional down regulation of hTERT and senescence induction in HepG2 cells by chelidonine. World J Gastroenterol 2009;15:3603–10.

[171]   Liu C, Liu Z, Li M, Li X, Wong Y-S, Ngai S-M, et al. Enhancement of auranofin-induced apoptosis in MCF-7 human breast cells by selenocystine, a synergistic inhibitor of thioredoxin reductase. PLoS One 2013;8:1–14.

[172]   Fadda AA, Amine MS, Arief MMH, Farahat EK. Novel synthesis, antimicrobial evaluation, and reactivity of dehydroacetic acid with N,C-nucleophiles. Pharmacologia 2014;1:1–11.

[173]   Liu S, Ng AK, Xu R, Wei J, Tan CM, Yanga Y, et al. Antibacterial action of dispersed single-walled carbon nanotubes on Escherichia coli and Bacillus subtilis investigated by atomic force microscopy. Nanoscale 2010;2:2744–50.

[174]   Ramírez-Lepe M, Ramírez-Suero M. Biological control of mosquito larvae by Bacillus thuringiensis subsp. israelensis Perveen F, editor. Insecticides: pest engineering. Croatia: In Tech; 2012. p. 239–64.

[175]   Tsang KW, Shum DK, Chan S, Ng P, Mak J, Leung R, et al. Pseudomonas aeruginosa adherence to human basement membrane collagen in vitro. Eur Respir J 2003;21:932–8.

[176] Patange VN, Arbad BR. Synthesis, spectral, thermal and biological studies of transition metal complexes of 4-hydroxy-3-[3-(4-hydroxyphenyl)-acryloyl]-6-methyl-2H-pyran-2-one. J Serb Chem Soc 2011;76:1237–46.

[177] Kannan S, Sivagamasundari M, Ramesh R, Liu Y. Ruthenium(II) carbonyl complexes of dehydroacetic acid thiosemicarbazone: synthesis, structure, light emission and biological activity. J Org Chem 2008;693:2251–7.

[178] Ullah H, Hamid Wattoo F, Wattoo MHS, Gulfraz M, Tirmizi SA, Ata S, et al. Synthesis, spectroscopic characterization and antibacterial activities of three Schiff bases derived from dehydroacetic acid with various substituted anilines. Turk J Biochem 2012;37:386–91.

[179] Batra N, Devi J. Synthesis of metal based chemotherapeutic agents derived from chloro-acetic acid [1-(4-hydroxy-6-methyl-2-oxo-2H-pyran-3-yl)-ethylidene]-hydrazide. J Chem Pharm Res 2015;7:183–9.

[180] Shaikh NP, Shaikh SF, Salunke SD. Synthesis and antimicrobial activity of substituted 3-cinnamoyl-4-hydroxy-6-methyl-2-pyrones. Int J Pharm Chem 2013;3:82–6.

[181] Munde AS, Shelke VA, Jadhav SM, Kirdant AS, Vaidya SR, Shankarwar SG, et al. Synthesis, characterization and antimicrobial activities of some transition metal complexes of biologi-cally active asymmetrical tetradentate ligands. Adv Appl Sci Res 2012;3:175–82.

[182] Shelke VA, Jadhav SM, Shankarwar SG, Chondhekar TK. Synthesis, spectroscopic char-acterization and antimicrobial activities of some rare earth metal complexes of biologically active asymmetrical tetradentate ligand. J Chem Sci Technol 2013;2:61–9.

[183] Siddiqui ZN, Praveen S, Musthafa TNM. Synthesis and antibacterial evaluation of novel heterocycles from 5-chloro-3- methyl-1-phenylpyrazole-4-carbaldehyde. Indian J Chem 2011;50:910–7.

[184] Borde VL, Naglolkar BB, Shankarwar SG, Shankarwar AG. Synthesis and characterization of antimicrobial activities of some transition metal complexes of asymmetrical tetradentate ligands. Res J Chem Sci 2015;5:19–23.

[185] Borde VL, Thakur CD, Shankarwar SG, Shankarwar AG. Synthesis and characterization of some Cr (III), Fe(III) and Co (II) complexes of unsymmetrical tetradentate schiff base. J Med Chem Drug Discov 2015 ISSN: 2347–9027.

[186] Kaur N, Aggarwal AK, Sharma N, Choudhary B. Synthesis and in-vitro antimicrobial activ-ity of pyrimidine derivatives. Int J Pharm Sci Drug Res 2012;4:199–204.

[187] Hamad Al-Obaidi O. Synthesis and spectral study, theoretical evaluation of new binuclear Mn (II), Ni (II) and Cu (II) metal complexes derived from pyrane-2-one and its biological activity. J Chem Bio Phy Sci Sec 2013;4:031–7.

[188] Saini S, Pal R, Gupta AK, Beniwal V. Microwave assisted synthesis and antibacterial study of hydrazone Schiff's base 2-cyano-N′-(1-(4-hydroxy-6-methyl-2-oxo-2H-pyran-3-yl) ethylidene)acetohydrazide and its transition metal complexes. Der Pharma Chem 2014;6:330–4.

[189] Chávez-Díaz IF, Angoa-Pérez V, López-Díaz S, Velázquez-Del Valle MG, Hernández-Lauzardo AN. Antagonistic bacteria with potential for biocontrol on Rhizopus stolonifer obtained from blackberry fruits. Fruits 2014;69:41–6.

[190] Durakovi L, Skelin A, Sikora S, Delas F, Mrkonji-Fuka M, Hui-Babi K, et al. Impact of new synthesized analogues of dehydroacetic acid on growth rate and vomitoxin accumulation by Fusarium graminearum under different temperatures in maize hybrid. Afr J Biotechnol 2011;10:10798–10810.

[191] Voigt CA, Schafer W, Salomon S. A secreted lipase of Fusarium graminearum is a virulence factor required for infection of cereals. Plant J 2005;42:364–75.

[192] Hertz-Fowler C, Pain A. Specialist fungi, versatile genomes. Nat Rev 2007;5:332–3.

[193] Walther G, Pawłowska J, Alastruey-Izquierdo A, Wrzosek M, Rodriguez-Tudela JL, Dolatabadi S, et al. DNA barcoding in Mucorales: an inventory of biodiversity. Persoonia 2013;30:11–47.

[194] Zadah ME, Jafari AA, Sedighi S, Seifati SM. Effects of dehydroacetic acid and ozonated water on *Aspergillus flavus* colonization and aflatoxin b1 accumulation in pistachios. Int J Pharm Ther 2016;7:90–6.

[195] Jadhav SM, Shelke VA, Shankarwar SG, Munde AS, Chondhekar TK. Synthesis, spectral, thermal, potentiometric and antimicrobial studies of transition metal complexes of tridentate ligand. J Saudi Chem Soc 2011;18:27–34.

[196] Durakovic L, Blazinkov M, Sikora S, Delas F, Skelin A, Tudic A, et al. A study of antifungal and antiafl atoxigenic action of newly synthesized analogues of dehydroacetic acid. Croat J Food Technol Biotechnol Nutr 2010;5:127–35.

[197] Plodpai P, Petcharat V, Chuenchit S, Chakthong S, Joycharat N, Voravuthikunchai SP. Desmos chinensis: a new candidate as natural antifungicide to control rice diseases. Ind Crops Prod 2013;42:324–31.

[198] Shaikh NP, Shaikh SF, Salunke SD. Synthesis, characterization and antimicrobial activity of new chalcones of 3-acetyl 4-hydroxy-6-methyl-2H-pyran-2-one. Int J Chem Pharm Sci 2013;4:97–101.

[199] Al-Jibouri MN, Musa TM, Hamad Al-Obaidi O. Theoretical and biological studies of binuclear Mn(II), Ni(II), Cu(II) and Cd(II) complexes with polydentate ligand derived from dehydroacetic acid and diethylenetriamine. Eur Chem Bull 2014;3:530–6.

[200] Winter CA, Risley EA, Nuss GW. Carrageenin-induced edema in hind paw of the rat as an assay for antiinflammatory drugs. Exp Biol Med 1962;111:544–7.

[201] Kumar A, Lohan P, Aneja DK, Gupta GK, Kaushik D, Prakash O. Design, saynthesis, computational and biological evaluation of some new hydrazino derivatives of DHA and pyrano-pyrazoles. Eur J Med Chem 2012;50:81–9.

[202] Mahdieh M, Yazdani M, Mahdieh S. The high potential of Pelargonium roseum plant for phytoremediation of heavy metals. Environ Monit Assess 2013;185:7877–81.

[203] Dias LC, Demuner AJ, Valente VMM, Barbosa LCA, Martins FT, Doriguetto AC, et al. Preparation of achiral and chiral (E)-enaminopyran-2,4-diones and their phytotoxic activity. J Agric Food Chem 2009;57:1399–405.

[204] Ezeja MI, Omeh YS, Ezeigbo II, Ekechukwu A. Evaluation of the analgesic activity of the methanolic stem bark extract of *Dialium guineense* (Wild). Ann Med Health Sci Res 2011;1(1):55–62.

[205] Gupta AK, Pal R, Beniwal V. Novel dehydroacetic acid based hydrazone Schiff's base metal complexes of first transition series: synthesis and biological evaluation study. World J Pharm Sci 2015;4:990–1008.

[206] Pal R, Kumar V, Gupta AK, Beniwal V. Synthesis, characterization and DNA photocleavage study of a novel dehydroacetic acid based hydrazone Schiff's base and its metal complexes. Med Chem Res 2014;23:3327–35.

Index

Printed in the United States
By Bookmasters